The Russell Cosmogony

A NEW CONCEPT
OF
LIGHT, MATTER AND ENERGY

by

WALTER RUSSELL

●

EPILOGUE BY

LAO RUSSELL

●

SPECIAL EDITION

Martino Fine Books
Eastford, CT
2023

Martino Fine Books
P.O. Box 913,
Eastford, CT 06242 USA

ISBN 978-1-68422-834-8

Copyright 2023

Martino Fine Books

Cover Design Tiziana Matarazzo

Printed in the United States of America On 100% Acid-Free Paper

The Russell Cosmogony

A NEW CONCEPT
OF
LIGHT, MATTER AND ENERGY

by

WALTER RUSSELL

●

EPILOGUE BY

LAO RUSSELL

●

SPECIAL EDITION

THE WALTER RUSSELL FOUNDATION

SWANNANOA

WAYNESBORO, VIRGINIA

WALTER RUSSELL

Author of

THE SECRET OF LIGHT

THE MESSAGE OF THE DIVINE ILIAD—Vol. I.

THE MESSAGE OF THE DIVINE ILIAD—Vol. II.

THE BOOK OF EARLY WHISPERINGS

SCIENTIFIC ANSWER TO SEX PROMISCUITY

YOUR DAY AND NIGHT

WALTER AND LAO RUSSELL

Authors of

UNIVERSAL LAW, NATURAL SCIENCE AND PHILOSOPHY
(A one year Study Course)

SCIENTIFIC ANSWER TO HUMAN RELATIONS

WALTER RUSSELL

DEDICATION

To my deeply illumined wife Lao I dedicate this book with a heart filled with gratitude for her guiding wisdom and selfless, indefatigable work during the last six years to make this presentation possible.

As the New Age of Transmutation slowly unfolds its new world for man may Lao's pervading genius be felt in these surviving words of the millions which have had to be destroyed when working alone without her Light in them.

Perchance the enduring lesson of my beloved Lao's life and mine is a demonstration of the infinitely multiplied power which comes to every man and woman whom God has fully joined together in spirit, giving to them their inheritance of His kingdom of the Light which they thus find through each other.

WALTER RUSSELL

An Open Letter To The World Of Science

Gentlemen:

This Open Letter to the World of Science, accompanied by a Treatise on The Russell Cosmogony, is being sent to approximately 350 members of our National Academy of Science, and Royal Society of London, 100 Universities and 300 leading newspapers.

This announcement with its new concept of Light, Matter, Energy, Electricity and Magnetism is a simple yet complete, consistent and workable cosmogony, which will enable future scientists to visualize the universe as ONE WHOLE, and will open the door to the New Age of Transmutation.

Recalling the important contributions I have already made to science, such as my work in completing the hydrogen octave and my prior discovery of the existence of the two atom bomb elements given to the scientific world in my two Periodic Tables of the Elements, assures me that you will give serious thought and attention to these documents.

Present threatening world conditions make it imperative that science discloses the way whereby the weakest of nations can protect itself from the strongest of them and render attack by land, sea or air *impotent.*

This new knowledge will give science this power.

England could have been rendered immune from her devastating bombardment had the world been receptive to these new scientific discoveries, which I endeavored to give to it when World War II started. Science however, did make use of the two atom bomb elements mentioned above, which I charted and copyrighted in 1926.

The world needs new metals. Many new rustless metals of greater density, malleability and conductivity await division in vast quantities from carbon and silicon. *These will be found when science discards its concept of matter as being substance, and becomes aware of the gyroscopic control of motion which will split the carbon tone into isotopes, as a musical tone is split into sharps and flats.*

In the chemial elements the sharps and flats are isotopes. These can be produced by man in greater numbers than Nature has produced them, for Nature does not begin to split her tones until she has passed two octaves beyond carbon. *There is a tremendous opportunity for the metallurgist of tomorrow to create new metals in the carbon and silicon octaves.*

Of even greater importance to the world in this crucial period is the production of unlimited quantities of free hydrogen. *This ideal weightless fuel could be transmuted from the atmosphere while in transit without the necessity of storage capacity.*

These are the important things which might now be known if Kepler's discovery had divulged the facts of geometric symmetry and dual curvature within the wave field.

His law of elliptical orbits evidences that he was on the verge of discovering that four—not two—magnetic poles control the dual opposed balance of this two-way universe. With but two magnetic poles a three dimensional radial universe of time intervals and sequences would be impossible. A balanced universe must have two poles to control centripetal, generoactive force, and two compensating poles to control centrifugal, radioactive force.

By means of such knowledge science could rid the earth of fear of attack by any nation, no matter how the attack might come, whether by land, sea or air.

This new knowledge will give to science the CAUSE of all the EFFECTS which have for centuries of research deceived the senses of scientific observers.

Man has a MIND as well as having senses, but he has given preference to the evidence of his senses in the building of his cosmogony. Man can REASON with his senses but he cannot KNOW with them. Reasoning is sense thinking—not Mind knowing. He has also produced EFFECTS without knowing their CAUSE.

The senses have not revealed to man that this is a *substanceless universe of motion only.* Neither have they told him the principle of polarity which divides the universal equilibrium into pairs of oppositely conditioned mates to create a sex-divided electric two-way universe.

The time has come in the history of man when KNOWLEDGE alone can save the human race. Man has for too long left the Creator out of His Creation, thinking He cannot be proven in the laboratory.

God not only can be proven in the laboratory, but because of the facts of that proof man can solve many heretofore hidden mysteries of the universe—such as that of the seed and growth—life and death cycles—the purpose of the inert gases as electric recorders of all repetitive effects—and the true process of atomic structure.

You might reasonably ask why I have withheld this knowledge for so many years. I have not withheld it. I tried in vain to give it from 1926 when I first published charts of the complete periodic table herewith attached, up to the beginning of World War II when I tried to organize a laboratory group to save England from its unnecessary bombardment.

I also accepted and held the Presidency of The Society of Arts and Sciences in New York for seven years for the sole purpose of giving to the world this new cosmogony based upon a two-way continuous, balanced universe to replace the one-way discontinuous, unbalanced universe, which is presumably expanding to a heat death.

During this period I lectured upon the misconceived idea that hydrogen is the basic number one atom of the periodic table. I explained that there are twenty-one other elements which precede it and that hydrogen itself is not a single element but a whole complex octave. I also explained the impossibility of there being any element without an inert gas as its source. At that time I distributed my periodic charts to approximately 800 scientists and universities.

Further than inciting research which yielded so-called isotopes of hydrogen and heavy water, nothing came of my effort, nor did I receive the credit due me. Incidentally those so-called isotopes are not isotopes but full toned elements of an orderly octave group series. Isotopes do not occur in Nature until they reach the octave following the silicon octave. The reasons for this are fully explained in our Study Course.

I wrote two books, gave many lectures and set up a demonstration laboratory in a university to prove that the elements are not different substances but are differently conditioned pressures of motion—and that the structure of the atom is based upon the gyroscopic principle.

As one after another of my discoveries appeared under other names, I acted on the advice of a friendly science editor to withhold any more of my new cosmogony until it was fully completed in words and diagrams, and again copyrighted.

It has taken many years to so complete it that it is invulnerable to attack, but this has now been done, and this present treatise is as complete in brief as the whole cosmogony is complete in detail.

I do not look for immediate acceptance of this revolutionary new knowledge. I do hope and expect however, that the seed of it will grow within the consciousness of science, and as I am nearing 82 years of age I feel it incumbent to announce the fact to science through this open letter and treatise, that The Russell Cosmogony, which my gifted wife Lao and I have together written into a year's Study Course of 935 uncontradictable pages and 182 diagrams, is now complete.

This course is now being studied all over the world and through our students as seed this new knowledge will ultimately transform the world.

It is with the deep desire that a higher civilization shall arise that I send forth this message to mankind. The day is HERE when Science and Religion *must* marry or through ignorance of God's Universal Laws man will perish from the earth.

Hoping that the world of science will recognize that this treatise has within it the answer to basic CAUSE for which it has been so long and tirelessly searching, I am

Sincerely yours,

Walter Russell

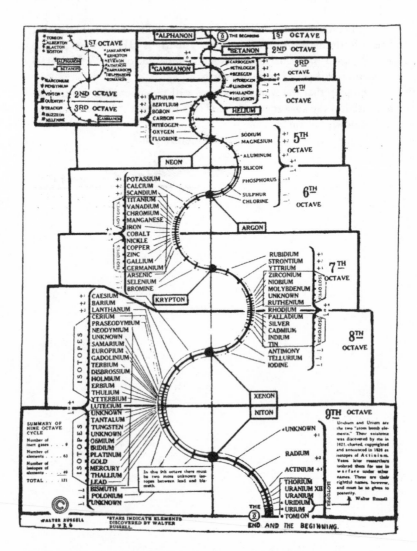

Figure 176. The Russell Periodic Chart of the Elements, No. 1

Periodicity is a characteristic of all phenomena of nature

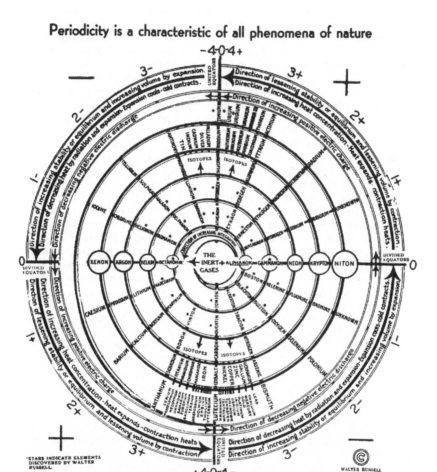

The nine octaves of the elements of matter manifest the polarization principle for producing dynamic action by extending two equators from a fulcrum point of rest. These two equators arise by gyroscopic action, multiplied centripetally, in four concentrative efforts to an amplitude plane which is 90 degrees from the zero plane of the inert gases. They then descend in four decentrative, depolarizing stages to disappear in their inert gases and again reappear from them in endless cycles throughtout eternity. Thus do all bodies appear and disappear — to again reappear—forever.

Figure 177. The Russell Periodic Chart of the Elements, No. 2

ACKNOWLEDGMENTS

There are many whom I have met on the long road to whom I owe much for open-minded interest, constructive help and sympathetic understanding. To these many friends I wish to express my gratitude for helping to smooth many rough spots on a seemingly impossible road, and for throwing just a little more light upon some of its dark intervals.

I therefore thank Dr. Henry Norris Russell for checking my first astronomical charts in 1922—Dr. George Pegram for warning me of the impossibility of ever hoping to force such a radical change in scientific thinking—Dr. H. H. Sheldon for placing a laboratory at my disposal at The New York University to demonstrate my hydrogen discoveries— the Westinghouse Lamp Company for giving me full use of its facilities for my gas transmutations, including their spectrum analyses—and the many who urged and aided hydrogen research which resulted in isolating several of those hydrogen octave elements shown on my new charts which were improperly named "hydrogen isotopes."

I feel especially indebted to the late A. Cressy Morrison for his vision and deep belief in my principles which he demonstrated by separating oxygen from nitrogen, and caused The Union Carbide Company to change its basis for producing hydrogen from coal gas instead of the electrolytic process—and to the late Thomas Edison for his more than passing interest in my ideas of polarity and the nature of electricity during my months of professional association with him as his sculptural biographer.

Many others to whom I owe my gratitude are Dr. Robert Andrews Millikan, Dr. Harlow Shapley, Dr. Willis D. Whitney, the late Doctors Lee de Forest, Nicola Tesla, Michael Pupin, Harvey Rentschler, and A. A. Michaelson, and Charles Kettering, David Sarnoff and Gerard Swope.

To the New York Times I also express my appreciation for the generous space given for the many letters for and against my teachings during my activities in the early thirties, and for naming my cosmogony "The Russell Two-Way Universe."

I gladly include in my appreciation those distinguished science writers William L. Laurence, Waldemar Kaempffert, John O'Neil, Gobindi Behari Lal and the late Howard Blakeslee, whose attitude toward a cosmogony so unlike that to which their training had accustomed them was always generous and sympathetic.

The attitude of all men of science with whom I have ever discussed my principles has always been cooperative, and I have met many during my seven years Presidency of The Society of Arts and Sciences, for intolerance is the usual reaction of human nature to any radical change.

WALTER RUSSELL

A Brief Treatise On
THE RUSSELL COSMOGONY

by WALTER RUSSELL

Once in a while, in long century periods, some vast new knowledge comes to the slowly unfolding race of man through cosmically inspired geniuses or men of super-vision, who have an awareness of the REALITY which lies beyond this universe of illusion.

Such new knowledge is of such a revolutionary nature in its time of coming that whole systems of thought, even unto entire cosmogonies, are rendered obsolete.

When each cosmic messenger gives such new inspired knowledge to the world the whole human race rises one more step on that long ladder of unfolding, which reaches from the jungle of man's beginnings unto the high heavens of ultimate complete Cosmic Consciousness and awareness of unity with God.

Thus it is that man has ever been transformed by the "renewing of his mind" with new knowledge given to him since his early beginnings, through the Mahabharata and Bhagavad-Gita of the early Brahmic days, through such ancient mystics as Laotze, Confucius, Zoroaster, Buddha, Plato, Aristotle, Socrates, Epictetus, Euclid, Mohammed, Moses, Isaiah, and Jesus, whose cosmic knowledge utterly transformed the practice of human relations of their day.

Then dawned a new day of the gathering of so-called "empirical knowledge," which is gained through the senses by research and observation of effects of matter in motion, rather than through the Consciousness of inspired Mind in meditation, which is the way that mystics and geniuses acquire their knowledge.

Since the days of Galileo this undependable method of gaining knowledge through the senses has served to multiply man's reasoning powers by teaching him HOW to do marvelous things with electricity and the elements of matter, but not one great savant of science can tell the WHY—or the CAUSE—of his familiar effects.

If asked what electricity, light, magnetism, matter or energy is he frankly answers: "I do not know."

If science actually does not know the WHY—or WHAT—or

CAUSE—of these essentials it necessarily follows that it is admittedly, without knowledge.

It is merely *informed*—but information gathered through the senses is not knowledge. The senses sense only EFFECTS. Knowledge is confined to the CAUSE of EFFECTS.

The senses are limited to but a small range of perception of the EFFECTS which they sense, and even that small range is saturated with the deceptions and distortions created by the illusion of motion.

It is impossible for the senses to penetrate any EFFECT to ascertain its CAUSE for the cause of illusion is not within effect. For this reason the entire mass of so-called empirical knowledge which science has gained by reasoning through the senses is invalid.

Let us examine some of these conclusions which form the basis of scientific theory and see why all present theory is invalid, and why its entire structure has no resemblance to either Nature's laws or its processes. I will now enumerate some of these unnatural theories.

BASIC MISCONCEPTIONS OF SCIENCE

1. *The cardinal error of science lies in shutting the Creator out of His Creation.*

This one basic error topples the whole structure, for out of it all of the other misconceptions of light, matter, energy, electricity, magnetism and atomic structure have grown.

If science knew what LIGHT actually IS, instead of the waves and corpuscles of incandescent suns which science now thinks it is, a new civilization would arise from that one fact alone.

Light is not waves which travel at 186,000 miles per second, which science says it is,—nor does light travel at all.

Science excluded God from its consideration because of the supposition that God could not be proved to exist by laboratory methods.

This decision is unfortunate for God IS provable by laboratory methods. The locatable motionless Light which man mistakenly calls magnetism is the invisible, but familiar Light which God IS—and with it He controls His universe—as we shall see.

MISCONCEPT OF ENERGY

2. Failure to recognize that this universal body of moving matter has been created by *some power outside of itself* has led science to con-

clude that the energy which created matter is *within itself*. Even more erroneous is the conclusion that energy is a *condition* of matter, such as heat.

This fallacy has led to the conclusion that Creation will disappear when heat energy "runs down." The first and second laws of thermodynamics are built upon this obviously wrong conclusion. The universe will never "run down." It is as eternal as God is eternal.

This universe of matter in motion is a Mind conceived, Mind creating body. As such it is as much a product of Mind as a pair of shoes, a poem, a symphony or a tunnel under a mountain is a product of the Mind which conceived it, and motivated the action which produced it as a formed body of matter.

The poem is not the poet however, nor is the symphony its composer. In a like sense this universe is not its own Creator. Whatever qualities, or attributes there are in any product—whether it be an adding machine or a universe—have been extended to that product by their creator to manifest qualities, attributes and energies, which are alone in the creator of that product.

Nor is the IDEA which matter manifests within matter. *IDEA is never created*. Idea is a Mind quality. Idea never leaves the omniscient Light of Mind. Idea is but simulated by matter in motion.

IDEA never leaves its invisible state to become visible matter. Bodies which manifest IDEA are made in the image of their creator's imaginings.

Every creation, whether of God or man, is an extension of its creator. It is projected from him by a force which is within its creator and not in the projected product.

All of the knowledge, energy and method of creating any product are properties of Mind alone. There is no *knowledge, energy, life, truth, intelligence, substance or thought in the motion which matter is*.

MISCONCEPTION OF MATTER

3. Electric matter is but a mirror which reflects qualities outside itself to simulate those qualities within itself.

In the Mind of *any creator of any product* is the IDEA of the formed body which Mind desires to produce. Also the *knowledge, energy and method of production* are in the Mind of the creator of that product and NOT in the product. The architect does not say that the energy, idea, or construction methods are in the temple of his conceiving, nor should man say that they are in the temple of God's conceiving.

To thus claim that energy is a property of matter is to deprive The Creator of His omnipotence and omniscience. The entire universe MANIFESTS power, but the universe is NOT THE POWER which it manifests.

Not one particle of matter which constitutes the material body of any product can move of itself. It can move only through desire and command of the Mind of its creator.

The powers of attraction and repulsion which science mistakenly attributes to matter are electrical effects performing their one and only function of dividing an equilibrium into two opposing conditions, which extend equally from a dividing equator. The magnetic Light controls the balance of these two opposing conditions, which interchange two-ways in their endeavor to void their opposing conditions, but the stresses and strains which *seem* to make matter attract and repel matter are electric effects.

Electric effects of motion can be insulated from each other—but the magnetic Light of The Creator, which causes those effects, cannot be insulated from matter by matter.

All matter is electric. Electricity conditions all matter under the measured control of the ONE MAGNETIC LIGHT which forever balances the TWO electrically divided, conditioned lights of matter and space."

Divided matter strains to find balance in the zero of equilibrium from which it was divided. The senses of man are mightily deceived by the illusions of appearance, which cause him to conclude otherwise.

Newton's apple was not attracted to the ground by gravitation. The high potential condition of that solid apple sought a similar high potential condition. That is to say it "fell" toward the earth to fulfill Nature's law of like seeking like.

Had Newton sat with the apple for a week or two he would have seen that same apple "rise" unto the heavens as a low potential gas seeking a like low potential position to balance its electrically divided state. The "rising" of the decaying, expanding apple again fulfills Nature's law of like seeking like.

All polarizing bodies add to their densities and potentials. The apple which fell to the ground was a polarized body. All polarized bodies must reverse their polarities and depolarize. They then lose their densities and potentials. The depolarized apple returned to the zero of its beginning.

The Newtonian law is, in this respect invalid, for it accounts for but one half of the apple's growth-decay cycle. This is a two-way uni-verse of opposed effects of motion—not a one-way universe.

MISCONCEPTION OF SUBSTANCE IN MATTER

4. Sense of observation has led to the erroneous conclusion that there are 92 different substances of matter.

This universe is substanceless. It consists of motion only. Motion simulates substance by the control of its opposing wave pressures of motion which deceive the senses into seeing substance where motion alone is. The senses do not reach beyond the illusion of motion, nor do those who believe that they can gain knowledge of the secrets of this vast make-believe universe even faintly comprehend the unreality of this mirage of polarized light in motion, which they so firmly believe is real.

Motion is two-way, for all motion is caused by the division of an equilibrium, and its extension in two opposite directions, to create the two opposite conditions of pressures necessary to make motion imperative.

One of these two conditions of electric motion pulls inward toward a center to create a centripetal vortice to simulate gravity. On the other side of the dividing equator the other condition thrusts outward from a center to create a centrifugal vortice to simulate vacuity.

Moving waves of oppositely conditioned matter simulate substance, but there is no substance to the motion which simulates IDEA in matter. If a cobweb could move fast enough it would simulate a solid steel disc—and it could cut through steel. If such a thing could happen it would not be the "substance" of the cobweb which cut through the steel—it would be the motion which cut it.

Fast moving short waves simulate solids, while slow moving long waves simulate the gases of space which surround solids. Waves of motion are substanceless, however. They merely simulate substance.

Motion itself is controlled by the Mind of the Creator, Who uses it to express His desire for simulating IDEA of Mind by giving it a formed body. There is no other purpose for motion.

Desire in the Light of Mind for creative expression is the only energy in this universe. All motion is Mind motivated. All motion records Mind thoughts in matter.

THE SECRET OF THE AGES

Step by simple step I will briefly unfold the supreme mystery of all time to enable science to void the confusion which has arisen from inability to relate the reality of the invisible universe to its simulation of reality, which has so regrettably deceived the senses of observers for all time. I do this, not only for science, but for the great need of religion, which so sorely needs a God Who can be KNOWN by all men as ONE, to replace the many imagined concepts of God which have so disastrously disunited the human race.

No one, save the few mystics of long ages, has ever known God, or God's ways. Nor has mankind yet known the meaning of LOVE, upon which the universe is founded—nor of LIFE, which the electric universe simulates in never ending cycles—nor of CAUSE of the EFFECTS for which man so heavily pays in tears and anguish for his not knowing.

The long heralded peace which passeth understanding awaits for science to tear away the veil which has for so long hidden the face of the Creator. Religion can be united as ONE only by dispelling the ignorance which now cloaks the faith-and-belief-God of fear which has bred so many intolerant groups of unknowing men.

We speak familiarly about the spiritual, invisible Mind universe of the Creator, and we speak with equal familiarity about the "physical" universe of matter which we call Creation, but the world has not yet known either of them separately, nor their unity as one to sufficiently define either of them scientifically.

I will now do this as simply as possible, in order that the physicist of tomorrow can KNOW and COMPREHEND the universe as ONE WHOLE, instead of SENSING it as many separate parts, which he will never be able to fit together.

I.

THE UNDIVIDED LIGHT

The basis of Creation is the Light of the Mind which created it.

God is the Light of Mind. God's thinking Mind is all there is. Mind is universal. Mind of God and Mind of man are ONE.

This eternally creating universe, which is God's eternally renewing body, is the product of Mind knowing, expressed through Mind thinking.

In the Light of God's Mind is all knowledge. All knowledge means full knowing of The Creator's ONE IDEA which is manifested in His Creation.

The undivided and unconditioned Light of Mind is an eternal state of rest. That invisible Light of the spirit is the equilibrium of absolute balance and absolute stillness, which is the foundation of the divided and conditioned universe of motion.

In that Light there is no change, no variance of condition, no form and no motion. It is the zero universe of REALITY. In it are all of the *Mind qualities* of *knowledge, inspiration, power, love, truth, balance and law,* which are never created, but are *simulated in moving quantities* in the divided universe of moving waves which we call matter.

The Light of Mind is the zero fulcrum of the wave lever from which motion extends. Its zero condition is eternal.

The unfortunate error of science lies in assuming that the power which belongs solely to the fulcrum of Light at rest, is in the motion of the lever which simulates that power.

II.
THE DIVIDED LIGHT

In the Light of The Creator's Mind is DESIRE to dramatize His ONE IDEA by dividing its one unconditioned, unchanging unity of balance and rest into pairs of oppositely conditioned units, which must forever interchange with each other to seek balance and rest.

DESIRE then multiplies those pairs of units into an infinity of eternal repetitions to give formed bodies to The Creator's imaginings. All formed bodies are created "in His Image."

Through the expression of DESIRE in LIGHT this universal drama of CAUSE and EFFECT is created as the product of Mind *knowing* divided by Mind *thinking.*

CAUSE is eternally at rest in the balanced unity of the undivided Light. CAUSE IS ONE.

EFFECT is eternally in motion to seek balance and rest in the centering equilibrium of the two opposed lights of this divided universe, which it finds only to lose. EFFECT IS TWO.

The Light of CAUSE, divided into the two opposed lights of EFFECT, is the one sole occupation of Mind which we call THINK-ING.

Mind thinking sets divided idea into two-way opposed motion to produce the effect of simulating idea *by giving form to it.*

Formed bodies are but motion, however. They are not the IDEA which they simulate.

III.
THIS ELECTRIC UNIVERSE OF SIMULATED IDEA

Mind thinking is electric. Divided electric thought pulsations manifest creative desire in wave cycles of motion, which forever vibrate between the two electric thought conditions of CONCENTRATION and DECENTRATION.

Concentrative and decentrative sequences of electric thinking produce the opposed pressures of compression and expansion, which form solid bodies of motion surrounded by gaseous space in one wave pulsation, and reverse that order in the next.

Concentrative thinking is centripetal. It focuses to a point. It borns gravity. It "charges" by multiplying low potential into high, and cold into heat.

Decentrative thinking is centrifugal. It expands into space. It borns radiation. It "discharges" by dividing high potential into low, and heat into cold.

All motion is a continuous two-way journey in opposite directions between two destinations.

One destination is the apex of a cone in an incandescent center of gravity. At this point motion comes to rest and reverses its direction.

The other destination is the base of a cone encircling a cold evacuated center of radiation. At this point motion again comes to rest and reverses its direction from centrifugal to centripetal.

So long as the Creator's Mind divides His knowing by His thinking just so long will that two-way motion continue its sequences of cycles to record God's imaginings in forms of His imaginings. God being eternal, likewise His universe is eternal.

The belief of science that the universe had a beginning in some past remote period—as the result of some giant cataclysm—and will come to an end in some future remote period is due to not knowing that waves of motion are the thought waves of the Universal Thinker.

Also the belief of science that the universe is dying a heat death by the expansion of suns is due to not knowing that there are as many black evacuated holes in space for the reborning of suns, as there are compressed suns for the reborning of evacuated black holes. See figures 101-102.

Together the interchange between these two conditions constitute the heartbeat of the universe, and they are EQUAL. Being equal they are balanced and continuous, eternally.

The journey toward gravity simulates life and the opposite journey simulates death in the forever repeating cycles which together, in their continuity, simulate eternal life.

The two opposite pressure conditions which control the life-death cycles of all bodies are:—(a) the *negative condition* of expansion which thrusts outward, radially and spirally, from a centering zero of rest to form the low potential condition which constitutes "space," and:— (b) the positive condition of compression which pulls inward toward a centering zero of rest to form the compressed condition of gravity, which generates forming bodies into solids surrounded by space.

Desire of Mind expresses its desire through the electric process of thinking. Thinking divides IDEA into pairs of oppositely conditioned units of motion, which record a simulation of IDEA into thought forms.

Sir James Jeans has suggested the possibility that matter might be proven to be "pure thought." *Matter is not pure thought, but it is the electric record of thought.* Every electric wave is a recording instrument which is forever recording the form of thought in wave fields of matter.

All thought waves created anywhere in any wave field become universal by repeating them everywhere.

Thought waves of expanded and compressed states of motion are fashioned into moving patterns which simulate the forms of the Creator's imaginings. All formed bodies thus created are "made in His image."

This division of the undivided Light and its extension into oppositely conditioned states of motion is the basis of the *universal heartbeat* of pulsing thought waves, which seemingly divide the ONE WHOLE IDEA into many ones.

Interchange between oppositely conditioned pairs of thought recording units is expressed in waves of motion.

This is a thought wave universe. Thought waves are reproduced throughout the universe at the speed of 186,000 miles per second.

It is commonly believed that the incandescence of suns is Light. Incandescence simulates Light in this cinema universe of macrocosmic make-believe, but incandescence is not Light. It is but motion. Incandescence is merely the compressed half of the one divided pair of opposite conditions, which constitutes matter and space. The black vacuity of cold space constitutes the expanded half. Together these two are as much mates as male and female are mates. Each is equally essential to the other. Each finds balance in the other by voiding each other's unbalance.

These two conditions and directions of compression and expansion are necessary for the two-way interchange of motion, which performs the work of integrating and disintegrating the living-dying cycles of opposed motion which this electric universe is.

The incandescence of compressed matter and the black vacuity of expanded matter are the two opposite polar ends of Nature's "bar magnet." Nature does not make her bar magnets in the form of cylinders as man does. She makes them in the form of cones. In this radial universe no other form of motion than the spiral form of cones is possible. See figures 158-159-160.

This means that the negative end of Nature's "magnet" is tens of thousands of times larger in volume than the positive end, although the potentials of each end are equal. It also means that the equilibrium plane which divides Nature's "magnet" is curved, while that same plane in a cylindrical bar magnet is a flat plane of zero curvature.

IV.
THE COULOMB LAW MISCONCEPTION

The Coulomb law statement that opposites attract and likes repel is not true to Natural law.

Opposite conditions ARE opposite conditions. Likewise, they are opposite effects caused by each pulling in opposite directions. It is not logical to say that opposites fulfill any other office than to OPPOSE. Nor is it logical to say that opposing things attract each other.

In all this universe like conditions seek like conditions. Gases and vapors seek gases and vapors by rising to find them. Liquids and solids seek liquids and solids by falling toward them.

Radiating matter seeks a radiating condition in the outward direction of radiation. Gravitating matter seeks the inward radial direction of condensation to find its like condition.

Opposite poles of a bar magnet push away from each other as far as they can go. That is the very purpose of the electric current which divides the universal equilibrium. If opposite poles attracted each other they would have to be together in the middle, instead of pushing away from each other to the very ends.

When depolarization takes place the poles seem to draw closer together, but that is because of their lessening vitality. They still pull away from each other until devitalization is complete. When motion ceases the matter which it manifests ceases to be.

Scientific observers have been deceived by their senses into thinking that opposites attract each other because of seeing the north pole of one magnet pull toward the south pole of another magnet.

The fact that opposite polarities void each other when thus contacted has not been considered as a factor in the matter. It is a fact however, when two opposites are thus brought together by their seeming eagerness to contact each other, both polls cease to be. Each one has voided the other as completely as the chemical opposites, sodium and chlorine void each other and leave no trace of either one after that contact.

If the Coulomb law were valid it would not be possible to gather together one ounce of any one element.

V.
THIS ELECTRIC UNIVERSE OF SIMULATED ENERGY

In order to know more dynamically what electricity really is I will define it. I will then amplify my definition by example.

Electricity is an effect of strain, tension and resistance caused by the energy of desire in the Light of Mind, to divide and extend the balanced unity of the ONE still Light of Universal Mind into pairs of many divided units of thinking Mind.

When electric strains and tensions cease to oppose each other electricity ceases to be. Electricity is dual action-reaction. When dual actions-reactions cease to vibrate, electric effect is voided by the one universal condition of rest.

Sound vibrations of a harp string are an electrical effect. The electrical vibrations of sound are a division of undivided silence. When sound vibrations cease, silence has "swallowed them up" by voiding them.

The IDEA of the silent harp string note eternally exists. Electrical division into sound manifests the IDEA, but the IDEA belongs to silence, and to silence it returns for reborning again as a simulation of IDEA.

The two electric pressures formed by the division of the universal equilibrium have separate offices to fulfill. The negative pressure expands to create space by dividing potential and multiplying volume. Conversely, the positive pressure contracts to multiply potential into solids by dividing volume.

Electricity thus performs the "work" of the world by straining toward separateness and multiplicity of units and also by relaxing from such resisted strains and tensions until motion ceases its vibrations by withdrawing into the universal stillness.

The only "work" performed in this universe is the "work" caused by the strains and tensions of electrically divided matter in motion.

Matter moves only to seek rest and balance.

Matter neither repels nor attracts matter. All matter which is out of balance with its environment, volume for volume—or potential for potential—will move *only* to seek rest in an equipotential environment of equal volume displacement.

That is why air or ocean currents move, and for no other reason than to seek their lost equilibrium. And while they move they will perform "work,"—and the measure of their power to perform "work" is the measure of their unbalance.

Earth's tides are not "pulled" by the moon. Curvature in the pressures of their wave fields which control their balance is the cause of that. And that explains why tides are thrust away from the face of the earth opposite to that of the moon, as well as being thrust toward the moon on its near face.

When tides rise they will perform "work," and they will also perform "work" when they fall, but "work" will cease being performed the moment the motion of either rising or falling ceases.

Likewise, a waterfall will perform "work" while falling but not when waters cease to move.

A storage battery will perform "work" while being "charged" with increasingly high potential pressures which oppose each other, and it will perform an equal amount while it is discharging to seek the equilibrium pressure which will unite the divided two. When fully discharged it will cease performing "work" because it has found balance in its zero and can no longer move.

In a live electric battery, or in its chemical counterpart such as sodium and chlorine, there are three equators, the central dividing one being the fulcrum of the two extended ones. When the two extended equators of the live electric battery withdraw into their balancing one the battery is "dead." They have found their eternal stillness.

Likewise their chemical counterparts have ceased to exist as separate elements when they withdraw into their sodium chloride fulcrum. Even though sodium and chlorine have disappeared they still are, for they will as surely reappear as night will follow day.

To recharge the battery the one dividing equator has to be extended in opposite directions until there are again three before motion is possible. *Motion is then not only possible but imperative.*

The heartbeat of the universe is eternal. So long as the universal heartbeat continues, every divided pair and every unit of every divided pair, will reappear to express life as surely as it will again disappear in eternal repetitions to express death.

"Work" is not performed by the attraction of matter for matter, nor because of a condition of matter, such as heat, which is presumed to be energy. "Work" is performed solely because the electric current, which divides a motionless condition into two unbalanced conditions, sets up two oppositely straining tensions of unrest which must move to release those tensions.

VI.
THE DUALITY OF ELECTRIC EFFECT

No effect can be produced unless there is an equal opposite effect to work with it. Electrical workers are two, which pull in opposite directions to perform that effect called "work."

Effect is therefore two-way, just as "work" is performed two-ways.

The two electrical workers are like two men on opposite ends of a double saw which pull and thrust in opposite directions from opposite ends to perform the "work" of sawing down a tree.

Or they are like two compression and expansion ends of a piston which pull and thrust in opposite directions sequentially, to move, and to perform "work" while they move, in either opposed direction. Each end of the saw, or piston, is helpless without the other.

Heat for example, is one end of the cosmic piston. Cold is the other end. Just so long as these two conditions exist, the piston of interchanging motion will continue to expand and contract sequentially. When each has found equilibrium by voiding the other, motion will immediately cease and "work" can no longer be performed.

Science says that cold is less heat. One might as appropriately say that female is less male, or that south is less north.

Science says also that there is no compensating uphill flow of energy to balance its downhill flow. There is an uphill flow. Otherwise a downhill flow would be impossible.

Every wave is a compression-expansion pump. The whole universe is a giant pump. The two-way piston of the universal pump constitutes the universal heartbeat. *A one-way universe is as impossible as a one-way pump is impossible.*

The compressed condition of this universe is exactly equal to the expanded condition. The compressed condition is gravitation. The expanded condition is radiation. *Gravitation and radiation are equal opposites.* Each is helpless without the other. In fact, each condition is impossible to produce without simultaneously producing the other. Heat is the effect of multiplied resistance to the compression of gravitation. Cold is the effect of the opposite strain of resistance to evacuation, or emptiness, which results from the expansion of radiation.

There is as much cold in the great expanses of space as there is heat in the compressed suns in all of this universe.

There is not one ampere of difference between these two opposite conditions of the electric workers in the whole universe, nor is there one milligram of weight in it which is not balanced between the two. This universe of electric waves is divided into wave fields. Each wave field is equally divided by contraction of gravitation and expansion of radiation. The potential of solids in a wave field is equally compensated by the potential of space which surrounds the solids.

It is as impossible to unequalize these two conditions in any wave field, or produce either one of them separately without simultaneously producing the other, as it would be to polarize one end of a bar magnet without producing an equal pole of opposition at the other end.

This wave universe is divided into wave fields. Each wave field is an electric battery which is forever being charged by the centripetal polarizing power of gravitation and discharged by the centrifugal depolarizing power of radiation.

This process is a manifestation of the life-death, growth-decay principle which is ever present in every effect of motion in Nature, without exception. Together they constitute the electric action-reaction sequences without which there would be no universe.

It is not true to Nature therefore, to say that either heat, cold, compression, expansion, or any other expression of motion is energy.

If the power to *cause* motion is in the balanced state of rest, it necessarily follows that energy is in the stillness of rest, and not in motion, which is effect of cause.

The Mind of the Creator is the fulcrum from which the wave lever of Mind thinking extends to express the energy of creative Mind. Thought waves cannot therefore, be the energy which caused them to become thought waves.

Any lever is powerless without a fulcrum. The power to move lies in the fulcrum which never moves.

All motion starts from a point of rest, seeks a point of rest and returns in the reverse direction to its starting point of rest. Test this fact by throwing a ball in the air, breathing in and out, pulling a chain, or walking.

Electrical effects of motion are not energy. Matter in motion is a marionette on the end of two Mind controlled electric strings.

VII.
WHAT IS THE "WORK" OF THIS UNIVERSE?

The only "work" performed in this universe is the "work" of recording thought forms of Mind imaginings into positively *charging* bodies, which are expressing the vitalizing half of the life-death cycle of creating bodies—and into negatively *discharging* bodies, which are expressing the devitalizing other half of that cycle.

That is the only work there is to do in all Creation, for God records His concentrative-decentrative thinking in the electric actions-reactions of living-dying bodies which appear and disappear in sequential cycles.

Creation of bodies is the only work that man does. Every body created by God or man appears from invisible stillness and disappears into that same stillness of its source, to reappear, periodically, in life-death, growth-decay cycles forever.

All bodies manifest eternal IDEA by eternally repeating their manifestations of IDEA in continuous cycles, which have no beginnings or endings. To exemplify: *cold generates—generation contracts—contraction heats—heat radiates—radiation expands and expansion cools.*

Sound—for another example—is a body of interchanging motion which appears from silence and returns to it. The silent harp string is the fulcrum of energy from which the moving harpstring extends as a vibrating lever of motion to manifest the IDEA of a musical tone in life-death cycles.

VIII.
THIS POLARIZED, SEX-CONDITIONED, PULSING, THOUGHT WAVE UNIVERSE

Science has for years been searching for some simple underlying basic principle of motivation which is present in every effect of motion. Mathematicians have hoped to find it and reduce it to a basic formula. Physicists have sought for it in the hope of thus discovering the life principle.

Science has never found it, and never will find it so long as it is sought for in either matter or motion.

That elusive secret is to be found only in the zero Light of the universal equilibrium, which is the fulcrum of the sex-divided electrical universe of thought waves of two-way motion.

That forever hidden secret of the ages is the *divider* of the ONE zero into a seeming TWO extended zeros. And it is the multiplier of the TWO into countless TWOS.

The name of that great divider of rest into two-way motion is
POLARITY.

Polarity is the controller—the measurer—and the surveyor of
intensity of desire in Mind for the actions-reactions needed for creative expression.

Polarity extends its surveyed measure of desire from a zero point
of rest in the universal Light, to two extended zero points of rest where
motion reverses its *direction*, its *polarity* and its *condition*.

These two points of stillness where motion reverses from one opposite pressure condition to the other are what science calls magnetic
poles. *The office of magnetic poles is to balance, and control, all divided
motion in the universe.*

Every particle of matter in the universe, whether atom or giant
sun, is controlled by a still centering point of magnetic Light. The two
extended poles of that still Light measure the intensity of desire which
motivates those extensions from their source of energy in the still
Light.

Polarity vitalizes and devitalizes—charges and discharges—
gravitates and radiates—inbreathes and outbreathes—lives and dies—
appears and disappears—compresses and expands—heats and cools—
grows and decays—integrates and disintegrates—and solidifies and
vaporizes by its electric actions-reactions which divide the ONE into
countless pairs of separate ones.

When man breathes in he polarizes his body. He vitalizes it into
wakeful action and an awareness of sensation. He charges his body
with higher electrical potential. He manifests life.

When man breathes out he depolarizes his body. He devitalizes
it into sleepy inaction and lessening awareness of sensation. He discharges his body by lowering its potential. He manifests death.

IX.
POLARITY PERIODICITY

Nature is engaged in the making of but one form—the cube-
sphere—which means the same as though we said female-male of man

The sphere is the positive centering sun. The cube is the invisible
surrounding wave field. All matter is thus divided into positive solids
surrounded by negative space.

As matter begins its formation into spheres its first shape is disc-

like, for it begins as the base of a cone. In a series of efforts which
constitute the octave wave, the first disc-like effort gradually prolates
until the perfect sphere is formed at wave amplitude. *This is the process by means of which "matter emerges from space."*

During this process the balance poles which control all matter
move gradually toward the pole of rotation. When the sphere is perfected, as it finally is at carbon, the two poles coincide with the pole of
rotation and the equator of the perfected sphere is 90 degrees from
the wave's axis. Likewise, the wave field becomes a true cube. Likewise,
any element which has reached its true sphere status will crystallize as
a true cube. Likewise, any divided pairs of elements which unite as one
on wave amplitude—such as sodium and chlorine—will crystallize in
the true cube shape of its wave field.

Conversely, as true spheres oblate, the two balancing poles move
away from the pole of rotation and toward the wave axis, until depolarization is completed and magnetic poles disappear in the plane of the
wave axis. *This is the manner in which "space swallows up matter."*

The mechanics of this process of polarization and depolarization,
under the guiding control of *two pairs of magnetic poles* will be more
fully described later.

This process of polarization takes place with increasing intensity
for one half of every cycle, whether of one breath, the cycle of a day,
a year, or a lifetime.

A man of forty will have reached his fully polarized strength to
manifest life in the first half of his life-death cycle. Depolarization
then assumes control as polarity reverses at the wave amplitude of
man's life cycle. Devitalization then begins and from there on man
manifests the death half of the cycle.

This process takes place in every creating particle of matter or any
combination of particles, whether in man, ant, electron, or nebula.

As polarization *increases* in intensity, the strains and tensions set
up by the desire of opposites of polarity to pull away from each other
increase in their intensity. This fact is exactly the opposite effect from
the conclusion stated in the Coulomb law.

As polarization *decreases*, the strains and tensions of electric opposition relax, until polarity entirely disappears in the rest condition
of the equipotential plane of the wave axis. This fact should not be
interpreted as opposite polarities attracting each other, for depolariza-

tion means that the ability to oppose lessens as each pole voids the other in the rest condition, but they still pull away from each other until their end.

The entire process of polarization and depolarization of every action-reaction of Nature could well be described as a lever reaching out in opposite directions from its fulcrum until it could reach no farther, then reversing those directions and *unwillingly withdrawing into its fulcrum* where motion ceases to again begin, and again reverse.

X.
SO CALLED MAGNETIC LINES OF FORCE

One of the great illusions of Nature which have deceived scientific observers is the principle of curvature, which is everywhere present in ever changing effect in every wave field, and in wave fields within wave fields throughout the universe.

Wave fields are bounded by planes of zero curvature, which act as mirrors to reverse all radiation which reaches out to these wave field boundaries.

An example of such a plane of zero curvature is the equator of a bar magnet. Iron filings reaching out from either pole will curve gradually in the ever changing pressure gradients which surround the poles. Science mistakenly calls these curved lines magnetic lines of force. See figures 65, 66, 67, 173 and 174.

When these curved lines reach the equator which divides the two poles, they reverse and repeat their curvature as though reflected by a mirror.

There are no magnetic lines of force in Nature. These so called curved lines are the radii of the spheres and spheroids which constitute this radial universe of prolating and oblating matter.

Radiation is an electric effect. It is not magnetic. Pressures which surround spheres and spheroids vary greatly in their equi-potential pressure gradients. As radiation is maximum at solar or planetary equators and gravitation is maximum at their poles, the pressure gradients surrounding spheres or spheroids vary in their curvature to conform to these pressures.

Gravitation and radiation are both radial. Radii of either the inward direction of gravity, or the outward direction of radiation, cannot be projected through varying pressures without bending to conform to the varying densities of varying pressure gradients.

Just as a stick, when thrust into water seems to suddenly break at the dividing plane of the two different densities, so likewise do the radii of incoming and outgoing light rays seemingly bend gradually as pressures gradually become more or less dense.

This divided universe is curved. Its two opposed conditions of gravitation and radiation are oppositely curved. Each has a system of curvature of its own and each system is opposed to the other; for their purposes are opposed.

The system of gravity curvature is evidenced in spheroidal and ellipsoidal layers of equipotential pressure gradients which curve around gravity centers. The surface of the earth and Heaviside layers are good examples.

The curvature of gravitation is centripetal. It is controlled by the north-south magnetic poles. Its office is to extend bodies in motion from their wave axes to their wave amplitudes.

The system of radial curvature is evidenced in ellipsoidal layers of equipotential pressure gradients which extend radially away from gravity centers. Radial curvature has the same relation to the equators of suns and planets as gravity curvature has to their poles of rotation.

Good examples of radial curvature are the rings of Saturn, the Dumb-Bell Nebula, (figure 148) and the sun's corona.

The system of radial curvature is centrifugal. It is controlled by two as yet unknown magnetic poles which will be amply described later as east-west poles.

The interrelations of these two pairs of poles is more fully set forth in chapter XXI.

The entire matter of curvature is one of the many optical illusions which Nature is completely made up of. Curved pressure gradients act as lenses to bend radiating light outward as they pass through their concavity from an inward to an outward direction. The reverse takes place as gravitating rays pass through the convexity of light lenses from the outward to the inward direction.

Polarity surveys and measures these pressures, but electricity alone projects and retracts the light which causes these illusions. The supposition that magnetism is a mysterious force of some kind which attracts and repels has helped to build these wrong conclusions, which the senses have deceived observers into believing. See figures 77, 161, 170, 171, 172, 173 and 174.

In our Study Course we have very carefully and plainly diagrammed the principle of two-way curvature within wave fields, and the principle of zero curvature which bounds wave fields and insulates the effect of one wave field from every other one by a principle of reversals, so we must let this brief description suffice for the purpose of this treatise.

XI.
THE INADEQUATE LAW OF CONSERVATION OF ENERGY

The law says that "the amount of energy in the universe is constant."

That is true because energy is unchanging in the undivided Light at rest. But the scientific meaning back of that true law is not Nature's meaning.

Energy belongs to the invisible universe. It is extended into the visible universe of motion ONLY FROM A FULCRUM which is at rest. The energy however, does not pass beyond the fulcrum into matter, or condition of matter, or motion of matter. That which passes beyond rest into motion is an *expression of energy*—a *simulation of energy*—an EFFECT projected from a CAUSE to demonstrate what energy can do when projected into the illusions of motion.

Energy thus expressed might be likened unto the countless actions of a motion picture. The motion thus expressed simulates the energy, and the IDEA, which has been projected *from* an undivided mental Source *through* a divided electric wave source, *by the way of* a fulcrum zero upon which the wave oscillates.

It cannot be said therefore, that the energy simulated by the motion picture is in the picture, rather than in the Source of the picture.

Likewise, the same cannot be said of Nature's cosmic cinema motion picture of CAUSE and EFFECT, which the Master Playwright has projected upon the screen of space from the Light of His knowing through the light lenses of His electric thinking.

To prove that the scientific meaning of the truly stated law is not Nature's meaning, I will quote from a science text book which explains the meaning of the law as follows: "This law means * * * that if energy *appears* in one form it must have *disappeared* from another in corresponding amount."

The words "appears" and "disappears" indicate that energy is meant to be actually within the visible motion and not in its fulcrum.

XII.
THERMODYNAMIC MISCONCEPTION

If polarity had been rightly understood by science the thermo-dynamic laws and accepted principles would never have been written. Clausius originally wrote the second law of thermodynamics as follows: "It is impossible for a self-acting machine to convey heat from one body to another at a higher temperature."

This is not true, for Nature is constantly doing just that thing in every expression of gravity. Every cold body of rain, or snow, heats as it falls to earth and conveys that heat to lift the higher temperature of the earth to a still higher temperature.

Every cold body which is added by gravity to a larger body at higher temperature raises the temperature of both bodies by the added crushing, compressing weight of gravity.

It is the office of gravity to compress—and it is the office of compression to heat—and it is the office of heat to throw off its hot bodies so they may cool and return as contracting bodies to again heat.

Every cold body which approaches a larger body "charges" both bodies. *Charging bodies heat.* Conversely, every body which radiates from a larger body "discharges" both bodies. *Discharging bodies cool.*

The cold of space heats hot suns hotter by the way of their poles, and hot suns radiate their heat by the way of their equators to form cooling rings, which again heat to become hot planets.

The law is further explained by stating that an object will fall of its own accord from a higher to a lower level, but it will not rise of its own accord from a lower level to a higher level.

This also is not true. Everything which "falls" toward one of the two polarized conditions of matter must "rise" toward the other opposite condition. The interchange is equal. The apple which falls of its own accord rises of its own accord.

Water unites its particles into closer relationship in order to fall, then divides into more remote relationship in order to rise. See Fig. 160.

Everything which emerges from space by the way of gravity is "swallowed up" by space by the way of radiation.

This is as true of suns as it is of apples. Every sun which is projected into space by one swing of polarity's pendulum has its mate in a black, vacuous hole of equal potential on the other side of its wave axis, which is waiting to "swallow it up" when the pendulum reverses its swing.

The confusion of observers who conceived that law is due to their not knowing of the balance of Nature which polarity controls. They think of the apple as a heavy object which cannot rise *as a heavy object*.

The layman thinks objectively of an apple as a solid object, but the scientist should think of the apple as one fleeting part of a whole cycle. The solid apple is that part of its cycle which has condensed from a large volume to become a small volume located at the apex of its spiral cone cycle.

Scientists should think cyclically and not objectively. The apex of a cone which one objectively thinks of as an apple, will expand to become the base of a cone which will eventually spiral into its apex on the bough of some tree, to again become an apple.

This is not an objective universe. It is cyclic. Objectivity is but one stage of a cycle which is forever moving through many stages between the appearance and disappearance of what the senses interpret as objective.

The astronomer should likewise think that way of his suns and stars. It should be easier for the astronomer to think cyclically, than for the physicist, for he can see his cone apices expand into cone bases for rewinding into new suns, in the same manner that apples expand to become cone bases for rewinding into objective apple forms at cone apices.

Figure 131 is a good example of the way an astronomer should think in relation to all stellar bodies. Had Newton thought that way in relation to an apple he would not have written such an inadequate, unnatural and misleading law.

Scientists should not think of sequential effects only, but should also think of the simultaneous workings of all two-way expressions.

This is how Nature works. As the solid apple FALLS, an equal potential SIMULTANEOUSLY RISES. If I put my hand down in water an amount equal to that which is displaced by my hand simultaneously rises.

As the apple falls it simultaneously *"charges" the earth* and *"discharges" space*. When the apple rises it simultaneously "charges" space and "discharges" the earth.

The balance in potential between gravitating matter and radiating matter in every wave field is absolute. Scientific observers have never thought of it that way. They have not thought of space as being divided into definitely measured "compartments" such as wave fields.

The more evacuated space there is in a wave field the more solid is the matter which centers it and the greater the volume of space. As space cools more, and its wave fields expand more in volume, its central sun equally contracts and heats.

Polarity divides all of its effects EQUALLY. The volume of negative space may be thousands, or millions of times greater than the volume of its positive center, but their potentials are equal to the millionth of an ampere.

If this were not so the Kepler law which says that "equal areas in a radius vector are covered in equal time" would not be workable, nor would the cube ratios of acceleration and deceleration work out.

XXIII.
INADEQUACY AND FALLACY OF NEWTON'S THREE
LAWS AND ONE HYPOTHESIS

The Newton laws and hypothesis seemed to be a master statement of Nature's underlying principles. They have held their prestige with reverence for their validity for three hundred years, during which time the misconception that all matter attracts all other matter with mathematically measurable power, has been a fundamental of scientific thinking.

New enlightenment as to God's ways and processes will demonstrate that this belief is but one more of the many seemingly obvious facts of Nature which deceive the senses into forming wrong conclusions. The senses of man have too readily accepted these *simulations* of facts for *real* facts.

Newton's first law says: *"Every body tends to continue in its state of rest or uniform motion in a straight line unless it is acted upon by an outside force."*

This law was written to fit non-existent premises which were mistakenly presumed to exist. *A body cannot continue its state of rest because bodies at rest do not exist in Nature. Bodies are but waves of motion. When motion ceases bodies cease to be.* One might as appropriately refer to sound being at rest in silence, for sound is matter in motion, like all other bodies.

A body cannot continue its uniform motion in a straight line in this radial universe of curved pressure gradients. Such a phenomena is impossible. Likewise all bodies are continually being acted upon by two outside, opposed forces—not one intermittent force.

Every body in the universe is constantly in violent motion, even though it simulates rest. When motion ceases matter ceases. A cloud, floating motionlessly above the earth, is moving at a speed of 1,000 miles per hour as the earth rotates. It is also moving violently in all of its parts.

It is also moving in a curved line of direction, not a straight line, even though the force acting upon it is unchanging.

The same thing may be said of airplanes, planets, moons, or radio waves. They all follow the curvature of gravity pressure gradients because gravity is always curved.

Even the lead pencil lying motionlessly upon a table cannot simulate rest except through motion so violent that your entire house would be instantly destroyed if the dual forces which cause that motion suddenly withdrew their support of it.

I herewith offer the following laws which have meaning in Nature to replace this meaningless first law:

1. *All motion in this polarized, radial universe is curved, and all curvature is spiral.*

2. *Every body is the result of the exertion of two opposing strains which pull against each other in opposite radial directions to condition its attributes and determine its motion.*

3. *Every body is perpetually in motion until the strains of opposition which keep it in motion void each other in the universal zero of rest, into which all bodies disappear for reappearance in reversed polarity.*

Newton's third law says: *"To every action there is an equal and opposite reaction."*

This law is inadequate and incomplete, for it confuses the facts

which govern motion. Just what does it mean? Had it been written in
either of the two following ways the confusion as to its meaning would
disappear, but either one would still be incomplete.

1. To every action there is an equal and opposite *simultaneous* reaction.—or—2. To every action there is an equal and opposite *sequential* reaction.

The inference is that the latter meaning was Newton's intent.

To rewrite this law in conformity with Nature's processes New
ton's third law should read as follows:

"*Every action is simultaneously balanced by an equal and opposite
reaction, and is repeated sequentially in reversed polarity.*"

That the above may be better understood by the scientist
whose traditional training has fixed within his Consciousness the one-
way universe concept, it would be well for me to give a few familiar
examples of the fundamental principle of the two-way concept, which
operates in every action-reaction effect of motion without exception
anywhere.

1. An outward explosion compresses in advance of the direction of
the action and *simultaneously* evacuates in the opposite direction. The
following cycle is in reverse. The evacuated condition becomes a compressed one, and the compressed condition becomes an evacuated one.

No better example of polarity than the above could be cited. All
electric division of the undivisible equilibrium into pairs of opposite
conditions takes place in this manner—and the only two conditions in
the universe are the compressed plus, and the expanded minus conditions and their resultant effects of heat and cold—male and female—
positive and negative and other wave vibrating pairs of opposites.

2. The discharge of a revolver and its recoil are simultaneous.
The sequential reaction is in reverse. That which was a discharge becomes a charge—and their directions are reversed. That which was
evacuated becomes a force of gravity which compresses at its center
instead of around its circumference. The concavity of outward pressures
reverses to the convexity of inward pressures.

Curvature of every simultaneous action-reaction is the reverse
of the sequential one.

3. The discharge of every outbreathing body, whether man, or
sun, or electron, charges space by compressing it, and simultaneously

discharges the body by evacuating it. The sequential action-reaction of inbreathing reverses this process.

Likewise, evacuating bodies simultaneously compress, and compressing space simultaneously expands its wave field boundaries to balance each opposite with the other.

Every outward action is an explosion which forms rings of compression at its equator. These surround a large volume of expanded space. The sequential inward reaction of this action forms a center of gravity within the expanded volume as a nucleus for a forming sphere. Every sphere thus formed "explodes" radially to form rings of compression at its equator which again become sphere systems.

This underlying process of Nature is present in its every action-reaction. It is the very mechanics of the universal inbreathing-outbreathing process of Nature which motivates the heartbeat of the universe. It is the inside-out, outside-in turnings of space and matter which swallow each other up to born each other sequentially. This process is Nature's most conspicuous fundamental. *Repetition in Nature is due to this process.*

4. Every growing thing which unfolds from the invisible pattern of its seed into visible form, simultaneously refolds within its seed as an invisible record of the pattern of the unfolding form. The sequential reaction reverses the inward refolding direction and repeats the outward unfolding direction of growth.

Newton's hypothesis states as follows: *"Every particle of matter in the universe attracts every other particle with a force that varies directly as the product of the masses and inversely as the square of the distance."*

Science states that Newton's mathematics prove without question that matter attracts matter. This is not true. Mathematics may prove the measures and relationships of a mirage, but they do not prove the mirage to be that which it simulates.

As an example, mathematics may prove without question that railroad tracks seem to meet on the horizon—but they do not prove that the railroad tracks do meet there.

Likewise, Newton's mathematics may prove the rates of acceleration and deceleration of the opposing pressures of gravity and radiation, as masses move toward and recede from each other in their eagerness to find rest from strains and tensions of unbalance, but that does not prove that matter attracts matter. It only proves that matter SEEMS to attract matter, just as railroad tracks *seem* to meet upon the horizon.

It would be just as logical to assert that planets were attracted by their perihelions because it could be mathematically proven that all planets increase their speed as they approach their perihelions.

For Newton's third law to be valid it must apply to all motion, such as the orbits of planets, as well as to falling bodies, such as the apple, which is claimed to fall because it is attracted to the earth. It must have no exceptions, and it has many. Let us consider one of them.

When the apple falls toward a center of gravity Newton adds up the product of the two masses—apple and earth—and mathematically accounts for the rate of acceleration as the two "mutually attracting" bodies approach each other.

When a planet approaches its perihelion however, its speed increases just as the speed of the falling apple increases. Unlike the apple which is approaching another body however, the planet is but approaching an empty point in space where there is no other body to add to that of the planet, such as there is when planet and apple potentials can be totalled.

The planet accelerates however, WITHOUT having another body to attract it. Why has not this most obvious fact been observed long ago?

Let us look at this law from another point of view. Science has founded its cosmogenetic theory upon its belief that this is a one-way, discontinuous, non-compensating universe, in spite of the very obviously continuous, compensating two-way universe manifested in all of its effects.

Whenever Nature projects any wave lever from its still fulcrum she projects it two opposite ways—simultaneously—then reverses both by withdrawing both lever extensions back into their fulcrum. This fulfills her invariable law, which decrees that all motion is born from rest, seeks two opposing points of maturity to rest, then returns as death to the zero of its beginning for rebirth.

THE TWO WAYS OF LIFE AND DEATH

Nature projects—or extends its wave lever for one half of its cycle to manifest the life and growth principle. That is the process of polarization. Polarization vitalizes bodies by dividing their zero condition of rest and extending the divided pairs away from their zero equator as far as they can go. Polarization pulls inwardly in centripetal spirals. It contracts to create gravity.

Nature then withdraws its wave lever into its zero source to manifest the death and decay principle during the other half of every cycle. That is the process of depolarization. Depolarization devitalizes bodies by voiding the desire of the divided conditions to oppose each other. It relaxes the strains and tensions of electric opposition. Depolarization thrusts outwardly in centrifugal spirals. It expands to radiate every generated body back into the zero of its source in order that death may reverse its manifestation and reappear as life.

Matter appears when polarization divides an equilibrium zero into two zeros of opposite polarity. Matter then disappears when the divided and opposed poles unite to rest in the eternal one zero of the still Light from which all things appear, and into which they disappear for the purpose of reappearing in eternal sequences.

WHAT ARE LIFE AND DEATH?

That which man calls life in bodies is accelerative motion only— the centripetal motion of interchanging wave vibrations between two poles, which have been extended from the ONE of their source. It is the accelerative motion of centripetal force which generates and contracts.

Science has long been searching for the life principle in some germ of matter. It may as well cast nets into the sea to search for oxygen.

That which is called death is just the opposite half of the whole life cycle. It is the decelerative motion of centrifugal force which degenerates, decays and expands.

Even though all bodies are both living and dying in each breath sequence of their whole cycles, the generative force of polarization is stronger in the first half of the cycle. Conversely all dying bodies are living while they die, but the degenerative force of depolarization which devitalizes is the stronger in the second half.

Life is eternal. There is no death. Life is but simulated in matter by polarization-depolarization sequences as all idea of Mind is but simulated in thought waves of moving matter.

WE NOW RETURN TO NEWTON'S ONE WAY LAWS
AND ONE WAY MATHEMATICS

Newton's one way laws and hypothesis account for falling bodies which are within the same wave field, and as a consequence have weight in respect to their common centers of gravity. Falling bodies are polarizing bodies. They gain weight as they fall.

Newton's laws do not account however, for rising bodies which have reversed their polarities and lose weight as they rise.

Neither do they account for floating bodies, such as suns, planets or moons which center their own wave fields, and as a consequence have no weight in respect to any other body in the universe.

Apples do expand into gases however, and rise. And liquids do expand into vapors and rise. *Cycles do not end in gravity. That is but their half way point where they simultaneously reverse their every attribute.*

They reverse their directions, their potentials, their polarities, their densities, their spectrum colors and their *weight.* One attribute cannot be reversed without reversing all. The polarizing direction of gravity multiplies the power of all expressions of force while the depolarizing direction of radiation divides them all in equal but opposite ratios.

The attribute of attraction which Newton gives to falling bodies exploding inward toward gravity, should also apply to rising bodies exploding outward toward expanded space. To apply that truth we would have to say: *Every particle of matter in the universe repels every other particle with a force which varies inversely as the product of the masses and directly as the square of the distance.*

Can it be true therefore, that every particle of matter attracts every other particle of matter in the universe and also repels every other particle? How can either or both be true when each denies the truth of the other?

Matter neither attracts nor repels matter. Matter moves in two opposite directions for the sole purpose of simulating IDEA in formed bodies by dynamic action-reaction sequences, and then seeks rest in the Light of IDEA to reawaken desire for again simulating IDEA.

All motion is unbalanced. All motion is forever seeking rest from its unbalanced condition by seeking voidance of its motion.

XIV.
THE FALLACY OF NEWTON'S MATHEMATICS

Even though an astronomer might find a new planet by applying the mathematics of Newton, this fact does not prove the claim made for it. A noted example of the attempt to prove a false premise by equations—which have merit in them as equations but do not have the least factual merit—was Newton's attempt to prove mathematically that the moon would fall upon the earth if it were not for a mythical "initial impulse" which gave the moon just the right speed to keep it from falling upon the earth or from flying off at a tangent.

As a preliminary to what I intend to say upon this subject, the moon is not only not "falling" upon this planet but is slowly spiralling away from it. Furthermore, all planets in any solar system—and all moons of all planets—and all suns, planets and moons of every nebular system in the heavens—are all spiralling away from their primaries. This is Nature's method of preparing for rebirth. Water vapor rises from water for the same reason—to disintegrate.

Every body disintegrates after it has passed the maturity point which marks the generative half of its cycle, but disintegration and death are but preparations for regeneration into life.

Suns wind up centripetally to polarize. When they have become true spheres they unwind centrifugally to depolarize. To depolarize they throw off rings from their equators. Rings become planets which likewise throw off centrifugally spiralling rings, which become moons.

This is Nature's method of returning her polarized bodies to the zero of their source. *She divides her masses into expanding systems and this division continues until matter has been "swallowed up by space."*

Newton evidently did not know of this depolarizing principle of Nature. He assumed that the moon has weight in respect to the earth, just as a cannon ball has weight in respect to the earth.

Believing in the deceptive evidence of the senses, he calculated the speed of continued momentum needed to keep a cannon ball from either falling to the earth or from flying off at a tangent. On the assumption that the moon has weight in respect to the earth—just as the cannon ball has—he proved to the world that the moon would fall to the earth if it were not for the "initial impulse" which kept the moon from falling.

And that is the belief of science today, because of the belief that weight is a fixed property of matter instead of being an ever changing property of ever changing polarity.

MISCONCEPTION OF WEIGHT

All freely floating bodies which are in balance with their environment have no weight whatsoever.

The moment any potential is taken out of an environment of equal potential, electric strains and tensions are set up in the unbalanced mass thus removed, which measure the resistance to that removal.

A stone for example, is very much out of balance with an equal volume of air. The stone will fall to seek a like potential, not because it has weight as a property of itself, but because of the strains of electric polarity which divide the universal balance into equal and opposite unbalanced pairs, and insist upon keeping the universe in balance.

Science believes that a man who weighs 150 pounds while surrounded by air still weighs 150 pounds when surrounded by the entirely different pressures of water in which he floats. That is not a true concept.

When a man is surrounded by air he is out of balance with polarity which divides pressures equally. Electric tensions then act as elastic bands which are sufficiently stretched to register a strain of 150 pounds pull against the zero of his balance. When he is surrounded by water however, the pressures of displacement and replacement are equalized. Each is in balance with the other and weight disappears.

If weight were a fixed attribute of matter it would be unchangeable. It varies however, as the potentials of masses out of balance vary. A man weighs less as he ascends a mountain, and more in a deep pit. As water falls it compresses and gains in potential. As it rises it divides into vapors and loses potential. It therefore, weighs less. When its potential is equal in volume to the volume of potential displaced it floats as a cloud. It then has no weight.

And so it is with stars, suns, planets and moons. They are all freely floating bodies and have no weight in respect to any other body in the universe.

Each solid centers an oppositely polarized volume of space in a wave field in which each polarized condition is of equal potential but of unequal volume. Every balanced wave field is insulated from every other wave field by reversals of curvature, which are separately con-sidered in later pages.

The following definitions of weight will help to clarify the present misconception regarding it.

1. *Weight is the measure of unbalance between the two electric forces which polarize the universal equilibrium.*

2. *Weight is the sum of the difference between the two pressures which act on every mass.*

3. *Weight is the sum of the difference in electric potential between any mass and the volume it occupies.*

4. *Weight is the measure of the force which a body exerts in seeking its true potential.*

5. *Weight is the sum of the difference between the inward pull of gravitation and the outward thrust of radiation.*

REGARDING "INITIAL IMPULSE"

The movement of planets around their suns, and of moons around their planets, has always been an unexplained mystery. In ancient days, even up to the 16th century, it was commonly believed that angels pushed the planets around in their orbits.

It is today quite commonly believed that at the time of the creation of this universe an initial impulse was given to each planet and moon, which was just sufficient to keep each one forever moving around its primary.

The slightest understanding of the nature of the electric current, and its mechanical process of dividing one condition of eternal balance into two opposed conditions of unbalance, would dissipate such a belief.

There are countless billions of suns, earths and moons in the heavens. It could not just "happen" that each of these has just the right initial velocity to keep it in its orbit as a result of the primal cata-clysm which is credited with the birth of the universe. That would be too great a cosmic coincidence for acceptance by any reasoning person.

Also such a theory would not bear the weight of so great a disturb-ing factor of this belief as the fact that no speed of any solar body is constant, as it would have to be to give substance to such a claim. Every solar body is forever, and constantly, varying its speed around its pri-mary.

It varies it in each revolution by going faster for one half, and slower for the other half. It also varies it over its millions of years of motion by gradually slowing its speed of revolution and increasing its speed of rotation as it spirals away from its primary. During these periods not one of the billions of solar bodies ever goes fast enough to fly off at a tangent from its primary, nor slow enough to fall into it, which in any case could not happen regardless of speed.

In addition to the foregoing is the fact that there never has been a primal cataclysm which created the universe. Electricity does not work that way, and there is no other working force in this universe than the dual electric force.

Electricity expresses its dividing powers equally—and simultaneously. Electricity then grows her effects to maturity and takes them apart to repeat them sequentially.

Also there have been millions of generations of suns, just as there have been millions of generations of men. If the initial impulse theory has any merit that merit would not apply to descendants ten times ten billion generations removed.

Our moon is not one minute old in cosmic time. It could therefore, have no "initial impulse."

This is a radial universe and every center of gravity in every solar body is the apex of a conic section. Every satellite of every such body is a radial projection from the equator of its primary.

It first appears as a ring thrown off centrifugally from its parent's equator. The ring becomes a sphere which centers its own wave field within its "ancestor" wave fields, then continues its outward spiral journey for millions of years of ever slowing speed and ever changing potential to keep in balance with the ever changing potential of its wave field.

Our solar system is a good example. Consider Mercury as the latest extension of our sun. It is a very hot and very compressed planet which speeds around its primary in less than three months. When Mercury spirals out to where our earth is it will take four times as long to make one revolution and it will be about four times as large in volume, for it must gradually expand to keep in balance with the ever changing equipotential layers of the pressure gradient which reaches out from the sun into space.

When Mercury attains the position of Jupiter it will be many times larger and its period of revolution will be many years. Also its period of rotation will speed up as its period of revolution slows down, in order that centripetal and centrifugal effects of polarization will keep their balance with each other.

Likewise, the inner moon of Mars circles its primary every seven hours, while the outer moon takes thirty hours.

All orbits are elliptical for they are angular conic sections. Likewise, all are either centripetal or centrifugal spirals, because their paths are either in the direction of the apex or the base of a cone.

Contraction in the centripetal direction accounts for increase of speed as planets approach their perihelions, and expansion in the direction of a conical base accounts for decrease in speed of revolution of outer planets, and also for decrease in speed as planets approach their aphelions.

In view of all such very orderly periodicities and processes in the formation of material systems, it seems incredible that a formula such as Newton's hypothesis ever should be thought of as proof that matter attracts matter, or that "initial impulse" accounted for the speed of planetary revolution.

XV.
TWO AS YET UNKNOWN FACTS OF NATURE

While considering Newton's laws I would like to touch lightly upon two as yet unknown characteristics of Nature.

1. One of these is the fact that every simultaneous and sequential action-reaction is in reverse of the other, yet Nature never reverses its direction from the instant polarization begins. It then extends its dividing effects in two opposite directions—each the reverse of the other—until both of those extensions are voided in their zero of origin, even though they traverse the universe in so doing.

The illusion of reversal is so convincing that it seems incredible that it is not factual. The inward pull of gravity is in the reverse direction of the outward thrust of radiation. Clockwise spirals are the reverse of anti-clockwise spirals, yet each is born out of the other without a reversal of direction, even though the effect seems to be in reverse direction.

This must have been intuitively divined by Newton when he wrote his first law. The words *"continue * * * in a straight line"* have in them a suggestion of his intuitive understanding of that principle, which he was unable to express in the measure of his inspiration. In many such ways he gives evidence of the mystic in him.

2. The other characteristic is the strange effect of polarity which causes all actions-reactions in Nature to forever turn inside out and outside in. This illusory effect contributes to the simulation of reversal in Nature, giving to Nature the sequences of solid bodies of incandescent suns, surrounded by vacuous black holes of space, followed by the reverse effect of vacuous black holes of space, surrounded by tenuous rings of what had been incandescent suns.

This fact will be further explained later when discussing the gyroscope principle.

Comprehension of this two-way pulsing effect in Nature makes it more easy to comprehend the rhythmic heartbeat of the universal cosmic pump, for its pistons must continue the inbreathing-outbreathing sequences in every particle of matter in the universe to simulate the eternal life principle by eternal repetitions of life-death cycles.

Comprehension of this fact will also clarify the illusion which this cosmic cinema really is, and also make understandable the reason why "any happening anywhere happens everywhere."

Further clarification of these secrets of the invisible universe would unnecessarily lengthen this treatise, but before passing I will give a law which is valid in Nature. This new law which is fully explained in The Russell Cosmogony is as follows: *Every action-reaction in Nature is voided as it occurs, is recorded as it is voided, and repeated as it is recorded.*

This is a zero universe of seeming. Its seeming reality is but a mirage extended from two zeros upon the blank screen of space to create the illusion of reality in an unreal universe.

XVI.
INADEQUACY OF KEPLER'S FIRST LAW

The saving grace of Kepler's mathematics lies in the fact that he did not try to prove by them a premise, or conclusion which is not true, as Newton did by claiming that his equations proved that matter attracts matter, and that the moon is falling on the earth.

His laws are free from such claims and demonstrate to a high degree the orderliness of effects of strains and tensions in a wave field. If wave fields were not balanced in their polarity such laws would not work out in Nature as they do. It is because of the absolute equality of division of opposing pressures in every wave field that such laws are workable.

Kepler's first law reads as follows: *"Each planet moves around the sun in an ellipse, with the sun in one of its foci."*

This law is right as far as it goes, but there are two foci to every orbit, and each of them have equal power in determining the rates of acceleration and deceleration of speed.

Just as Newton's law accounted for the falling apple, but ignored the other half of the apple cycle from the zero of its beginning to the zero of its ending, so likewise, Kepler's first law accounts for but one half of an orbit by the reference to *only one of its two foci.*

Every orbit is balanced and controlled by four magnetic poles, not two. It has not yet been known that there are four magnetic poles, but a three dimensional cube-sphere universe would be impossible with only the two north-south poles. I will enlarge more upon these four magnetic poles in a later chapter, and describe the separate offices which each fulfills in the extension and retraction of wave fields.

The two unknown east-west poles which control the shortening and lengthening elliptical orbits are those inferentially referred to in the Kepler law.

In order to comprehend the periodicity of the familiar north-south magnetic poles, it is necessary to comprehend the relationship between these two opposing pairs of north-south and east-west magnetic poles, and their manner of extension from a common fulcrum and their retraction to it.

As planets oblate, their north-south poles gradually move away from their poles of rotation. Our earth has become sufficiently oblate for its magnetic poles to move to an angle of 23 degrees from the pole of rotation.

This periodicity is balanced by an angular periodicity of planets' equators moving away from the equators of their centering suns, where all planets of all systems are born from rings.

The equator of our earth has moved out of the plane of the solar equator to keep pace with the polar shift. Each must balance the other. The four magnetic poles control that balance. These are the important facts which should have been inquired into when Kepler wrote his law.

It is not important to know that the sun is in one of its orbital foci if the tremendous significance of the two foci is ignored. The amazing fact is that matter and space are playing seesaw with each other in the proportions of an ant and an elephant. The mechanics which balance and control such a stupendous "game," with such mathematical precision is the important thing to know.

An ant and an elephant can play seesaw if the ant has a sufficiently long lever, and that is practically what is happening throughout the universe—suns and planets being the ant and space being the elephant.

The planets and moons of all solar systems are gradually tearing their suns into rings in a most orderly manner, with a precision which is mathematically measurable in direct and inverse ratios. The four magnetic poles—in two opposing pairs—control this amazing performance of Nature within every wave.

That is the important thing to know. By means of this knowledge of God's ways in Nature we can make them our ways in the laboratory, and thus have a command over Nature, which man has never had beyond the comprehension of his day.

It seems incredible that Kepler could have known of these two east-west poles without having realized their purpose, and their necessity in a three dimensional universe.

Knowledge of God's ways will alone give science the power to balance all effects in the universe of man's making as God balances them in the universe of His making. The precision of every effect in God's universe is so perfectly managed that an astronomer can calculate to the split second the position of any planet in the solar system, or as accurately foretell an eclipse of one by the other.

The secret of man's ability to control his universe lies in the knowledge of the tonal octave wave and its field. Therefore know the wave in all of its simplicity of three times three in numbered effect, multiplied to infinite complexity but still not passing beyond the three times three of man's easy comprehension.

The first great step in acquiring knowledge of the wave field must be the great revolution in scientific thinking in regard to matter itself.

In order to control matter, science must know what it is and all of the various steps of its generation from zero into form, and its degeneration back to zero. It must know matter for what it IS instead of believing it to be what it is not and working with it on that premise.

The rest of this treatise will be devoted to clarifying the meaning of this one subject.

XVII.
REGARDING THE QUANTUM THEORY

This theory claims not only that energy is within matter, but that it exists in "bundles." Its very basis has no relation to Nature anywhere, nor to the workings of polarity—the great divider—nor to the electric wave.

One part of the theory describes certain microscopic "resonators" embedded within particles of matter to make it vibrate. These are set in motion, according to a recent article in Scientific American, by light entering through holes which must be of just the right size in every case, to cause the vibrations to release these "bundles" of energy. Nothing could be more fantastic nor more of a travesty of Nature, for the only *cause* of vibration is polarity.

The only vibrations there are in Nature are those interchanges between the two opposites of polarity which extend from a fulcrum zero to a plus and minus zero. These are the destination points between which motion oscillates.

XVIII.
REGARDING SINGLY CHARGED PARTICLES

Just as it is impossible to polarize the positive end of a bar magnet without simultaneously polarizing the negative—or to depolarize one end separately—or to create a battery of one cell without simultaneously creating its opposite cell—or to create one hemisphere of a planet without simultaneously creating the other—or to lift one end of a lever without simultaneously lowering the other—or to deep freeze without generating heat—so it is impossible for man or Nature to produce singly charged negative, positive or neutral particles.

There are no negatively "charged" particles in this universe. Negative electricity *discharges* while positive electricity *charges*. The negative depolarizing force functions in the opposite manner and direction to the positive polarizing force.

Positive electricity produces the condition of gravity by compression—which means charging or generating.

Negative electricity produces the condition of radiation by expanding—which means discharging or degenerating.

It is impossible for one of the polarized conditions to be present without the other, for each opposite borns its mate and interchanges with it until each one becomes the other.

All particles of matter in the universe are alike in one respect, whether that particle is an invisible electron, planet, or sun. That universal attribute is the fact that each has two opposing hemispheres which are under the control of two opposing balance poles. One pole controls its charge and the other its discharge. Together they keep the universe in balance.

As there is not one law for microscopic mass and another for colossal mass, let us consider the earth as a typical example, keeping in mind the fact that colossal mass is but many small particles. The earth is being constantly *charged* into higher potential by the centripetal multiplying force of positive electricity which polarizes and vitalizes. Incoming sun rays to earth are a good example. Conversely, the earth is being constantly *discharged* into lower potential by the centrifugal, dividing force of negative electricity which depolarizes and devitalizes. Witness outgoing earth rays.

Both are the same rays. They have but changed their polarities by reversing their outward direction of expansion to an inward direction of contraction. When sun rays leave their cathode in the sun they are negative particles—or vortices of motion which we call matter. Their polarity constantly changes until they change their direction at the equator between sun and earth. They then become centripetally contracting vortices instead of centrifugally expanding ones. After passing their equator their polarity is positive instead of negative. Their positive charges increase as they near their anode, the earth.

The very reverse effect takes place in respect to radiation leaving the earth which is now the cathode for the projected vortice of spiral motion, and the sun is its anode.

The simultaneous charge and discharge of every particle, or mass of particles, is repeated sequentially in wave pulsations which constitute the universal heartbeat. Every particle in the universe breathes in and out in polarization-depolarization sequences.

As there is no exception to this law it cannot be possible for Nature or man to create particles which are singly charged. Science has listed about twenty of these separately charged particles and claimed for them different attributes, just as the elements are presumed to be different "substances," with different attributes.

The time has come when we must think of matter in a new way. Our old concepts of substance—and of the attributes of substance, which we call matter, must radically change.

That revolutionary change is what I now wish to talk about. The preceding pages are but a preparation for a complete transformation of thought concerning matter.

XIX.
FUTURE SCIENCE MUST COMPLETELY REVOLUTIONIZE ITS CONCEPT OF MATTER

For ages man has thought of matter as being substance.
Posterity must learn to think of matter as motion only.

The senses of man have for such long aeons told him of the many different substances which compose the universe. Therefore it will not be easy for him to make this transition.

The granite rock, the iron bar, the steel ship, and the many other substances which hurt him by too rough a contact, or burn him with their heat—or refresh him with their cool wetness—or nourish him with the meat of their bodies—or lend their bodies to him for the fashioning of countless things of his desiring—all these many things of seeming substance of earths and seas have told his senses of his apparently substantial body.

They have told him that matter is substance—and that it is *real*. It unquestionably exists. Objectivity of matter is the most obvious fact of the universe to man's senses.

All down the ages the mystics have *affirmed* that the universe is but illusion, and that "there is no life, intelligence or substance in matter." Abstract affirmations however, are not convincing to either scientist or layman whose senses have taught him otherwise.

Unfortunately for the world, those who speak abstractly, and affirm without being able to exlain dynamically, have been listened to with ears which could not hear that which had no meaning for them.

The time has now come however, to give meaning to the inspired mystics and poets, who have been illumined with inner knowledge which they found impossible to put into words for man. The foundation principle of the universe is utterly simple, but the simplest of stories is the hardest to tell.

It will not be easy for either the layman or the scientist to make the transition in his thinking from a universe of real and dependable substance to a substanceless thought wave universe of motion, whose sole purpose is the recording of thought imaginings. The result of such motion is to create a make-believe universe in which both substance and form are simulated by as many states of motion as there are simulated substances and forms in matter.

The scientist has not only divided matter into 92 different kinds of substances, but he has divided these 92 substances into atomic systems made up of many more minute particles of somewhere around twenty "primal" substances. These he calls electrons, protons, neutrons, antineutrons, antiprotons, photons, gravitons, mesons, kappa mesons, positive mu mesons, negative mu mesons, positive pi mesons, negative pi mesons, neutral pi mesons, tau mesons, positive v-particles, negative v-particles, neutral v-particles and so on without an end as yet in sight of the many non-existent substances. They so convincingly act their parts in producing the mirages of substance in this universe, that the greatest scientists of this world have not the slightest suspicion that the many different substances of matter are but different states of motion.

The reason for this great confusion is because scientific observers started out with the wrong premise from the very beginning. With an unwarranted belief given to the evidence of their senses, science—ever since Democritus—has been searching for an irreducible unit of matter which would account for the universe.

It never seems to have occurred to any of the great thinkers of the ages that Creation could not create itself. Just as the picture does not paint itself but must have its source in the painter, or as the poem cannot write itself but must also have its creator poet, so likewise, must this master-drama of CAUSE and EFFECT have its Creator Playwright who conceived the IDEA of Creation and gave it form.

One might as reasonably scrape to the bottom of Leonardo da Vinci's painting, The Last Supper, to find its primal pigment and its first brush stroke in the hope of finding the IDEA which the painting manifests, and its creator as well.

As neither the IDEA of Creation, nor the Creator of it, are in the pigment which painted The Universal Masterpiece, and likewise, as the primal pigment and first brush stroke were not the Cause of the picture, one might search for countless ages without finding that for which he sought.

Having thus started out with the wrong premise regarding matter, and having passed on these wrong premises and assumptions through generations of teachings, which have become traditions to their inheritors, it is not strange that the conclusions of science are as invalid as the premises upon which they have been founded.

Science is still searching for the primordial life principle in matter as eagerly as it has searched for the primordial substance from which other substances extend.

The time has come in man's mental unfoldment when he must recognize that all IDEA is eternal in the zero equilibrium of the still magnetic Light of Universal Mind—which IS God—and that IDEA is but manifested in motion of body forms by polarized cycles. These APPEAR from the eternal zero and MUST DISAPPEAR into that zero in order that they may REAPPEAR in endless cycles.

The layman, as well as the scientist, must think differently and express his thoughts with a greater understanding of Natural Law. The layman for example, who says: "John is dead," and thinks of John as being John's body, is not expressing the facts of Natural Law.

John is not John's body. John is not dead, nor can he die. The eternal John is an IDEA of Mind. His depolarized body is but one end of a longer cycle than the cycle of his incoming-outgoing breath, but not one whit different. His breathing will repolarize in another second, but his depolarized body will take a longer period, just as the depolarized oak will repolarize again from its zero in its seed.

The scientist must also think in terms of polarity and use terms which have that connotation in them. I will give an example of my meaning by quoting a paragraph from The Scientific American, Jan. 1952:

"Even more confusing is the fact that all the mesons, pi and mu, undergo *spontaneous disintegration.*" Why not say *that they have become depolarized—or that they have attained their equilibrium—or that they have attained zero! The simple fact is that the motion which gave them seeming being has ceased.* There is no such effect in Nature as "spontaneous disintegration."

I will quote the rest of the same paragraph to call attention to the needless complexity of descriptive terminology which belief in substance seems to make necessary. It is as follows: "When left alone in free space, a positive or negative pi meson decays into a positive or negative mu meson and a neutrino within about 250-millionths of a second. Then the positive or negative mu meson decays into a positive or negative electron plus two neutrinos within about two-millionths of a second. The neutral pi meson also is unstable and decays into two gamma rays in a very short time indeed—about a hundred-millionth of a millionth of a second."

The above is a very complex and confusing way of saying that matter has disappeared by depolarization because motion has ceased.

God's universe consists solely of vibrating waves of two-way interchanging motion. Every effect in Nature is included in that simplicity. Any child will fully comprehend you when you tell him that sound is an effect caused by rapid vibrations. You can demonstrate it by plucking a harp string so that he can see that the sound is caused by rapid motion.

You do not even need to tell him that the sound ceases when the motion ceases. His common sense will tell him that.

If however, you tell him the wire is composed of positive protons which decay into negative electrons in a 100-millionth of a second to produce sound, then the negative electrons decay into silence in another hundred millionth of a second, he will look at you blankly and understand not one word of it.

All of these many named particles, which science thinks of as differently charged substances, are all basically the same spiral units of motion. These are constantly being transformed from one condition to another, as each divided pair obeys the polarizing charge of gravity until it has completed the outward half of its journey to its reversal point of rest. It then returns as each one depolarizes and withdraws within its fulcrum zero of rest.

Exactly the same thing is true of all of the elements. Science has given them 92 names and listed their many attributes, such as metals, metaloids and nonmetals—alkalis and acids—brittle and pliable—conductive and nonconductive—dense—liquid—soft—gaseous—and many other attributes.

To anyone, whether scientist or layman, a piece of iron, a piece of aluminum, and a lump of gold are three different metals which have always been and always will be just what they unquestionably are— three unalterably different substances.

Any other interpretation of them would be unthinkable.

That is the kind of thinking however, which must be relegated to past ages. *Mankind must henceforth learn to look upon matter as a transient motion picture record of the idea which it simulates. For that is what it really is—a Cosmic cinema thrown upon the majestic screen of space.*

XX.
THE NEW CONCEPT OF MATTER

All of the many seeming substances in this universe are but many different pressure conditions. These have been created by the interchange of two-way motion between two opposed poles of rest, which have been extended from the zero universe of KNOWING MIND to simulate the multiple ideas of THINKING MIND.

Any form of matter becomes another form of matter if its pressure condition is changed. Nature perpetually changes one form of matter into another by perpetually changing its pressure conditions.

Every element in the entire periodic table is a transmutation from the preceding element of its cycle, from its beginning in zero to the ending of the entire nine octaves in the zero of its beginning.

The age of transmutation of the elements by man begins when he has full knowledge of the manner in which Nature transmutes one element into another.

CREATION—POSTULATED PROGRESSIVELY

1. This universe of moving body forms is an expression of the desire for division of the formless, sexless, Father-Mother balanced unity into pairs of equally and oppositely unbalanced, disunited, sex-conditioned father and mother moving body forms.

2. The purpose of this division into sex-conditioned, disunited pairs of father and mother moving body forms is to eternally extend the desire for unifying disunited father and mother body forms in order to eternally extend desire in them for repeating their sequences of division and unity.

3. The only energy in the universe is the pulsing desire of Mind

for the creative expression of Mind knowing, by giving thought-imaged body forms to the IDEA of Mind knowing.

4. The only means of expressing the pulsing desire of Mind idea is through the concentrative-decentrative pulsations of Mind thinking.

5. Mind thinking is electric. The desire pulsations of electric thinking are concentrative and decentrative.

6. Concentrative thinking focuses idea into patterned form in seed of matter to manifest the fatherhood of Creation. To focus is to compress. The product of concentrative thinking is the compression of gravitation which fathers all body forms.

7. Decentrative thinking expands conceived idea from its patterned seed and extends it outward from seed idea to give it body form of idea, and thus manifest the motherhood of Creation. To extend is to expand. The product of decentrative thinking is the expansion of radiation which mothers all body forms.

8. The mother pole of Creation unfolds the moving body from its seed idea and projects it toward its zero in the heavens of space.

9. The father pole of Creation refolds the extending mother form into its seed and withdraws it toward its zero in body forms of earths.

10. All body forms of matter give forth pulsing life as its action and receive pulsing death as its reaction.

11. All body forms of matter are both womb and tomb of all life and death.

12. All life is born from death—and death is born from life—for reborning death and life. All opposites born each other and become each other in alternate sequences.

13. The seed is the fulcrum zero from which the divided father and mother body forms extend, and sequentially return for re-extension. The seed of all things centers all things. It is the fulcrum of the eternally manifested tree of life, and of every root, branch and leaf —and of every corpuscle of every root, branch and leaf.

14. The Soul centers the seed of all idea. All action-reaction pulsations of living-dying body forms are recorded in the Soul-seed of all living-dying body forms. All living body forms are dying as they live, and living as they die. Veritably, death is born in the very cradle of life, and the tomb again cradles death as life.

15. The electro-chemical records of the zero seed of all things are the zero elements which are known as the inert gases, from which

center of the fulcrum zero of polarity all polarizing body forms extend to manifest vitalizing life, and return as depolarizing forms to manifest devitalizing death.

16. The inert gases are God's recording and repeating system. They record, remember and repeat all actions-reactions of all things from eternity unto eternity. They broadcast all of Creation to all Creation and likewise receive the broadcasts of all Creation for rebroadcasting to all Creation.

17. The inert gases are zeros in the universal equilibrium. Polarity divides and extends the One Light into electric thought wave cycles, which appear from the One Still Light as pairs of moving lights and disappear into that still Light for reappearance forever without end.

18. The inert gases are the "spiritual elements" which born and reborn the physical elements, and meticulously make spectrum records of their eternities of rebornings.

19. The inert gases center all elements from within to control their unfolding cycles of polarizing-depolarizing form, and balance them from without by two poles of still Light to control their refolding of form into their zero seed.

20. The inert gases record purposeful unfoldings and give back to each corpuscle of motion its cell memory of purpose and its instinctive guidance.

21. They likewise give back to awakening Consciousness the records of all cycles of Soul awakening, which have been written in the Soul-seeds of all unfolding-refolding body forms.

22. The inert gases write down in God's books of Light all that John, and Bill, and Sue have ever been—likewise what the ant, the elephant, the tiger, violet and bee have ever been—or have ever done since their beginnings—and give them back to them after every rest period which divides their cycles.

23. God's sole "occupation" is the building of moving body forms to simulate His One Idea of CAUSE and EFFECT which Creation is.

All CAUSE lies within the unconditioned, balanced, magnetic Light of Mind knowing.

All EFFECT lies within the two unbalanced, polarized lights of electric thinking, which create the two unbalanced and opposed conditions which Creation is.

24. Electric thinking divides all effect into opposite pairs *equally*.

Each one of each pair of effects is equal. Their balance is absolute.

The balance of the universe cannot be upset by even one millionth of an electron's weight. The answer to this secret lies in further solving the mystery which surrounds polarity. Polarity has never been understood. It must now be understood.

25. *Question. How can there be motion in a balanced universe?* If two children of equal weight sit at opposite ends of a seesaw, or two equal weights are put on scales, there is no unbalance—but likewise, there is no motion. Unless there can be unbalance there can be no motion. How can there be unbalance in an equally divided and equally balanced universe?

Answer. Two children of equal weight, playing seesaw, do not interchange with each other while they are at rest. When they desire to move they throw themselves out of balance with their fulcrum by their equal leanings, but they are in balance with each other. Motion is then imperative. When thus thrown out of balance they must reverse their leanings to restore balance and lose it again, as all things in Nature do.

Nature has a different way of playing seesaw. Instead of oscillating upon a continually extended lever, the wave extensions of polarity withdraw into their fulcrums and re-extend by turning inside-out and outside-in.

Nature plays seesaw with matter and space as opposite mates. It is as though an ant and an elephant played the game. When they interchange, the ant swells to the elephant's volume and the elephant shrinks to the volume of the ant. Both are of equal potential however, for the solidity of one balances the tenuity of the other.

The cause of continued motion and sequential reversals lies in the two opposed conditions of matter. The compressed center heats and heat expands, while tenuous space cools and cold contracts. The necessary reversals of Nature's wave lever, because of difference in volume between the ant and elephant, produce the same effect by throwing the players out of balance with their fulcrum.

XXI.
THE UNKNOWN AND UNSUSPECTED MYSTERY
OF MAGNETIC POLES

26. *There are four magnetic poles in every wave field, not two, as*

heretofore believed. A three dimensional cube-bounded, sphere-centered, radial universe would be impossible with but two magnetic poles.

The two unsuspected magnetic poles are not unknown, however. They are the two foci so casually referred to in Kepler's law of elliptical orbits, and they are in a plane of 90 degrees from the plane of the positive and negative north and south poles.

The two as yet ignored magnetic poles have already been referred to as east and west magnetic poles. The office of these east and west positive and negative poles is to control the balance of prolating and oblating spheres and their orbits, as they contract into spheres and expand into rings equatorially, in opposition to the north and south poles, which control the balance of extension and contraction in the direction of rotating poles.

27. Nature is engaged solely in the manufacture of spheres of solid matter surrounded by cube wave fields of tenuous space. Spheres are created by extending the flat discs, which are the inert gases, into rings and spheroids which gradually become spheres. The opposition of the north and south magnetic poles is accountable for that. They pull away from each other as hard as they can to fulfill the generative half of the electric cycle.

The generative half is the polarizing half. It is the vitalizing half, comparable to the maturing years of a man's life from babyhood to forty years. The north and south poles pull not only against each other's resistance but against the opposite pull of the east and west poles, which finally conquer the generative power of gravity and oblate spheres into spheroids, then pull spheroids into rings and discs until the depolarization process is complete. The depolarizing radiative half of the cycle might be likened to the aging latter half of a man's life.

The forces of pulling and thrusting are electric. The division into opposite conditions is electric. Magnetic poles control and balance the two electric dividers of the universal equilibrium but the work of extension from the fulcrum of stillness is entirely electric.

Electricity is the engine which supplies the motivating force to the universal ship but polarity supplies the rudder, the gyroscope and the balance which every moving body must have.

Electricity is the physical expression which Creation is, but the magnetic Light of the universe is the Source of that expression which acts under the spiritual direction and control of magnetic poles of Light. Poles appear only when motion begins its division of ONE into TWO, and disappear when the TWO cease to be two in their unity as ONE.

28. Nature generates matter from rings into spheres by the way of north-south poles and radiates spheres back into rings by the way of their equatorial east-west poles. *In this manner matter emerges from space to form moving bodies, and is swallowed up by space to disappear into the stillness of their zero source.*

RECIPROCATIVE WORKINGS OF OPPOSING POLES

North-south poles balance and control the prolating of spheres which Nature needs for the forming of bodies and their division into pairs. They extend in opposite directions at angles of 90 degrees from wave axes to form poles of rotation for spherical body forms. They are the shafts of waves and of all spheres which spin upon shafts.

East-west poles balance and control the oblating of spheres which Nature no longer needs for its body forms. They remain on wave axes to extend equators for forming spheres. They are the rims of wheels which spin upon the north-south shafts.

North-south poles control the division of equilibrium into two opposite conditions which occupy opposite sides of mutual equators.

East-west poles exercise their control from equators of forming spheres and balance the movements of all orbits and all aphelions and perihelions of orbits as matter appears from its fulcrum and disappears into it.

East-west poles mark upon sphere's equators the *seeming* oscillations of the north-south piston strokes as the compression of gravity and the expansion of radiation cross and recross equators to perform the work of unfolding and refolding body forms of Mind idea.

North-south poles control centripetal windings of spheres which form where the apices of two cones meet, and east-west poles control centrifugal unwindings of spheres and sphere systems into cone bases at wave axes.

North-south poles divide the ONE condition into TWO, against the resistance of east-west polarity, while east-west poles unite the TWO conditions into ONE against the resistance of north-south polarity.

North-south polarity for example, controls the electric division of

the one balanced condition of sodium chloride into two unbalanced conditions. Sodium chloride is the fulcrum. Sodium and chloride are opposite ends of a lever which is extended from the fulcrum like two children on opposite ends of a seesaw.

East-west polarity controls the electric withdrawal of the two extensions into their fulcrum, thus uniting the two extended equators
with their fulcrum at wave amplitude. Instead of three equators for the
two extensions there is now but one equator for the united pair.

North-south poles give one of the three dimensions which this
dimensionless equilibrium needs for the projection of its illusions,
while east-west poles give the other two.

The one dimension of north-south polarity is length, for poles of
rotation have no other dimension as they are but one radius of a sphere.
The other two dimensions are width and breadth, for equators of
spheres are circles, and circles have infinite radii.

North-south poles extend away from each other at an angle of 90
degrees from their equators to divide the universal one condition into
two opposed conditions.

East-west poles remain upon the planes of their equators to unite
the two divided conditions into one balanced condition.

North-south directions lead away from each other, out into
infinity. They are opposites and opposites oppose. They never can become one but they can be *voided*.

East-west directions lead into each other to void all polarity.

XXII.
THE ILLUSION OF THREE DIMENSIONS
AND HOW THEY APPEAR

29. The electric action-reaction of universal thinking might be
likened unto an outward-inward explosion. This Mind universe is
engaged in thought expression everywhere. From every point in the
universe little and big outward-inward, polarizing-depolarizing explos-
ions are continuously taking place.

The outward actions manifest the giving half of the cycle of the
Love principle which motivates this universe. The inward reactions
manifest the regiving half of the cycle. Nature never takes. It but gives
for regiving.

An action anywhere is repeated everywhere. The measure of

desire for action is measured out on wave axes in octave harmonics at a speed of 186,000 miles per second. Octave harmonics on wave axes are east-west magnetic poles. The same measure of desire is marked out from the same zero source in the north-south polar directions, which extend from the centering zero at 90 degrees from the equatorial plane of the east-west poles.

Matter is born at zero degrees. Polarization builds it up to maturity at 90 degrees. Depolarization then returns it to the zero of its birth.

If similar balloons were inflated they would touch each other at six points on six curved surfaces. To continue the inflation until the empty spaces were filled, would flatten those six curved surfaces until they became six flat planes of zero curvature.

That is what happens in Nature. Cube wave fields are thus formed to bound wave fields, and to insulate one from another by compelling a reversal of direction and polarity when radii meet those planes of zero curvature.

The entire inner structure of every wave field is curved, beginning with the sphere which centers it, and ending at the planes of zero curvature which bound it.

30. Every wave field is a cosmic projector which radiates light outward through the concave lenses of spheroidal pressure gradients, to bend toward the mirrors of wave field boundaries of zero curvature, where curvature reverses as it is reflected into neighboring wave fields. It is also a receiver of light rays which bend inwardly toward its center of gravity by way of the convex lenses of pressure gradients.

31. True cube wave fields occur only where true spheres are formed. This occurs in only one place in the entire nine octave wave cycle. That one place is carbon. The crystals of pure carbon are true cubes. I will amplify this fact later.

32. The three dimensional illusion of Nature is caused by a series of three light mirrors of zero curvature which center the cube in three planes—all of which are at right angles to each other—and six boundary mirrors of the cube—which are likewise at right angles to each other. See Figures 106 to 114.

33. *This is a zero universe of rest from which motion is projected into seeming existence, and then is retracted into its zero of rest.*

That zero bounded field of reversed motion withdraws within its central zero as it depolarizes, leaving a complete record of the pattern of its actions-reactions in the zero inert gas of its octave wave, for repolarizing into the same patterned form as it reappears.

34. Every action-reaction is three. Three is the basic number of this universe. Three is a two-way polar extension of its centering source. Three is the fulcrum and the lever. Three is the expansion-contraction from a centering source of your heartbeat and the heartbeat of the universe.

Three is balance extended to two equal and opposite balances. Three is the sexless Father-Mother, divided and extended to the sex-conditioned father and mother. Three is your inbreathing-outbreathing, and it is the piston of the wave-trough-wave-crest, compression-expansion pump which this universe is.

Three is the one dimension of polarity—north and south, or east and west—but the three dimensioned volume which polarity centers and bounds is three multiplied by three.

Three is the sphere, for the sphere is but one form of the three dimensions of length, breadth and thickness. Its radii are alike in all three dimensions. It has no diagonals, angles or planes.

Nine is the hot spherical sun crystallized into the cold cube of space. The cube is nine dimensional. Its eight tones and fulcrum are nine. Its eight diagonals and fulcrum are nine. Its three extended planes and six boundary planes are nine.

Nine is the octave wave which consists of four extended pairs, centered by the zero of their source.

35. Beyond nine Nature cannot pass. Every action-reaction however, must add up to nine. Not one event in Nature can be more, or less than nine. See figure 114.

XXIII.
THE EARTH IS NOT A MAGNET

36. It is commonly stated in science text books that the earth is a giant magnet. That is not true to Nature's processes. The equator of a magnet is not a center of gravity. The center of the earth is a center of gravity.

All matter, whether of earths, suns, or corpuscles, is formed between the opposite poles of two magnets. To produce the effect of gravity, two dividing equators must be united as one. See figures 78 to 83.

Man's bar magnets are cylinders of unchanging condition. Nature's magnets are cones of ever changing conditions. See figures 158-159.

The equator of man's magnets is of zero curvature and centers its poles. The equator of Nature's magnets is curved and is off center. Much confusion has arisen from this misconception. See figures 159 and 160.

XXIV.
EVERY PARTICLE OF MATTER IS BOTH CATHODE AND ANODE

37. This is a radial universe of ever changing pressures. Every extending particle which leaves a cathode or anode is negative, for it expands as it leaves its primary and thus discharges. That very same negative particle—electron or otherwise—changes its polarization intensity every millionth of an inch from either its cathode or anode. That is the reason science has so many names for the same particles.

When a particle arrives at wave amplitude—or any equator where pressure condition is reversed—it can well be called a neutron, for its polarity is balanced at that reversal point.

After its curvature is reversed it then becomes a positively charging particle, for it contracts as it radially approaches its anode. It might then be called a positron or positive meson, or many other names as its condition changes. See figure 77.

XXV
THERE ARE NO SEPARATE PARTICLES OR ELEMENTS

38. This same principle applies to all of the elements of matter. All of them are made up of the same units of opposed motion. We call them hydrogen, iron, carbon, sulphur, magnesium, nickel and many other names. We think of them as separate substances having separate properties.

All of the elements are made up of the very selfsame particles—which should more appropriately be called units of motion—or vortices. The only reason we have for thinking of them as different substances is because they have certain predictable effects upon each other and upon our senses.

The fact is however, that their pressure conditions are different in every part of the wave in which they find themselves. Lithium particles become boron particles when the gyroscopic relation of the plane of lithium's orbit changes to the plane occupied by boron—and so on during the whole nine octaves of changing pressure conditions.

XXVI.
CURVATURE IS ALSO POLARIZED

39. There are two opposed systems of curvature within wave fields. One system of curvature produces spheroidal pressure gradients around north and south poles, which makes of each pole a gravitative and radiative center.

The other system of curvature produces spheroidal pressure gradients around east and west poles. These cause gravitative and radiative centers to form between north-south poles and unite divided pairs into one.

Pressure gradients act like lenses in the bending of radii inward or outward, in accord with the direction of motion through convex or concave curvature. This is too vast a subject for further expansion here, but diagrams from The Russell Cosmogony Course, figures 162 to 174 inclusive, will partially explain what is more fully explained in the Study Course.

XXVII.
EVERY CONDITION OF MATTER IS DEPENDENT UPON ITS OPPOSITE CONDITION

40. Polarity creates moving body forms in pairs of opposites, and places the opposite of each pair on opposite sides of a mutual equator. It likewise makes each mate so dependent on the other that neither one could survive without constant interchange.

No living body could survive without receiving its inward breath from its spatial counterpart, nor could the spatial mate survive withou' the outbreathing of its opposite body to recharge it. Nor could any living thing exist without interchanging with its root mate. Neither tree which is attached to its roots, nor man who is not attached but equally depends upon interchange with them, could exist without the other.

The dry cell could not maintain its condition without interchange with its opposite on the other side of the equator. To charge one charges both alike. To discharge one means death for both alike. Every sun has its equal and opposite mate in a black vacuous hole on the other side of its equator. The sun is compressed into greater vitality by the coldness which feeds it from its expanded black hole mate. Conversely, the black hole is expanded into greater vitality by the heat which feeds it from the sun; thus the cycle is completed by this interchange.

Sequentially the black vacuous hole will become the sun and the sun will become a black vacuous hole. See figure 102.

XXVIII.
VIBRATING MATTER—THE ROOT PRINCIPLE OF ATOMIC STRUCURE

41. The A string of a harp is silent because it is not in motion. If you pluck it you will set it in motion. By doing so you have created a material body of sound by producing motion. The sound body which you have created by the energy of desire extended from your Mind into matter is the tone of A.

The reason you have produced the tone of A is because the wire was conditioned to produce that tone.

If you tighten the wire, thus changing its condition, you produce the tone of B. By repeatedly contracting the wire by tightening it you can produce a whole octave of increasingly higher tones. Conversely, a whole octave of lowering tones can be produced by expanding the wire by loosening it.

The same wire can produce many tonal sounds. It produces different tones by changing the condition of the wire. The change is a difference o compression or expansion.

42. The birth of any body from its fulcrum zero to its zero of maturity and back again to its fulcrum is a cycle. Cycles of wave vibrations in an electric current, or in a musical note where the vibrations are so fast that the sound is heard as a continuous tone, are known as wave frequencies.

The high frequencies of the tone of a harp string are too fast for the ear to hear the GROWTH of that tone from its birth to its maturity and back again. The ear hears only the fully grown matured tone repeated in hundreds of cycles in one second without being aware of the growth of hundreds of bodies through whole life cycles from birth to death.

For this reason we do not think of a sound as a body as we think of a man, tree or bird as being a body. It is a body, however—just as a man is a body—for it is motion, and all motion is material body.

We have many years to witness the many changes in a man's life cycle. We even have periods of many years each of witnessing different stages of that growth, such as the childhood period—then the boy— the young man—the mature man—then the aging man—and the very old man.

Every creating body progresses through these periods of growth from birth to death, whether that body is one hundred millionth of a second cycle of a high frequency electric current—an eighty year cycle of a man—or a million billion year cycle of a sun. There is no difference except in time.

43. One can better comprehend the meaning of this idea by taking a slow motion sound film of the harp string tone of A. By so doing we lengthen the life cycle of the sound body to one cycle in sixty seconds instead of one cycle in a three or four hundredth part of a second.

Now you can witness the growth of that tonal body from its "infancy" to its "old age." You will first hear a faint low sound, which is no more like the full grown tone of A than a baby is like a full grown man. Gradually it grows through its early stages toward maturity and the sound you hear is like a siren growing ever more shrill until the fully grown tone of A is reached at its maturity.

If you could see that body it would be a true sphere in form, compressed into a very small space at the very middle of the extended harp string. It would be the center of gravity for the wave field, created by dividing the equilibrium of silence into a conditioned state of motion. The rest of the wave field would be "empty space" of millions of times greater volume.

Its pole of rotation and the axis connecting the two magnetic north-south poles, would coincide and would be parallel to the harp string vibrating axis. Its equator would be 90 degrees from that axis.

Gradually you would see that sphere flatten and throw off rings from its equator—and you would hear that siren tone in reverse until you could no longer hear it. Motion has ceased by withdrawing into its silent source. We do not say it is dead, for we know it will be repeated on the other side of its equator instantly. We say that a man is dead because the time interval between repetitive cycles is so long that we do not realize that the law of repetitive vibrations is the same for all cycles, nor do we realize that all cycles are alike in the fact that they grow to maturity and die, in order that they may be reborn to again live and die.

The space which surrounded the sound sphere has "swallowed it up." The sun turns inside out to become space and space turns outside in to become the sun on the other side of the wave axis. You again hear the siren growth of the tone of A.

44. If we now "speed up" the cycles of man to as many frequencies as the tone of A, all we could see of his cycle would be the matured man. We could not see the childhood, boyhood or manhood stages of his cycle. Instead of seeing the matured man of one cycle, we would see hundreds of matured cycles of that same man, without being able to see the changing stages of any of the cycles.

45. If we likewise "speed up" the growing cycles of the stars in the heavens, we would see them come and go like fireflies flashing in the meadow. Their tens of billions of years duration is just a difference in the timing of their cycles, but the principle of growth and decay is identical in every state of motion created to manifest every idea in Creation.

The answer to that is polarity. Polarity is expressed in waves. Waves have dimension. Time is a dimension. It takes time to create a wave, for a wave is a cycle which has a seeming beginning and ending. The time cycles for reproduction of electric thought waves is constant, but the time periods of life-death cycles vary as thought waves accumulate into cycles upon cycles of formed bodies of sound, or of insects, animals, men, trees, suns or nebulas.

Herein lies the solution of the mystery of growth and decay, or life and death, which has been deemed insoluble during all the ages of man.

Life and growth are thought waves multiplied by time—while decay and death are time divided into voidance.

This whole universe is but a projection of Mind Idea into a three dimensional universe of thought wave timed sequences from the uni-

versal zero of rest, followed by a voidance of that projection by withdrawal within the universal zero of rest.

XXIX.
THE MYSTERY OF GROWTH AND DECAY— AND OF LIFE AND DEATH

46. Man conceives the idea of life and death of his body as a beginning and ending of the idea of himself. Back of that concept is the belief that his body is himself.

There is no beginning or ending of any effect in Nature for there is no beginning or ending of cause. CAUSE is eternal. EFFECT is eternally repeated.

Man's body is an eternally repeated effect of its cause, which is eternal man. Man is an IDEA—a part of the ONE WHOLE ETERNAL IDEA. Idea is unchanging. Bodies alone change. Idea is never created. Bodies alone are created to manifest idea.

All bodies are sequential repetitions of effects. All effects in Nature "rise" from the zero of eternal rest to manifest IDEA through action. They do this for a period of time, then they lie down to rest before again going into action. *There is no exception to this principle in all the universe of mighty stars and microscopic particles.*

All actions of all bodies are always under the control of Mind which caused them. Bodies have no power to move through their own initiative, for they have no energy or initiative of their own. Initiative is extended to bodies by the universal Mind which controls them.

Even though this treatise is for the purpose of explaining the mechanics and processes made use of by Mind to create matter, *we must not for a moment forget the reality of Mind nor the illusion of matter.*

In continuing therefore, to explain Nature's methods of unfolding bodies from their Soul-seed idea into form, and refolding the records of those forms into their Soul-seed idea, *we should cultivate the realization that we are dealing with thought wave patterns of idea, and not with substance or matter.*

THE MYSTERY OF TIME

47. When we think of matter we should think of the thought waves which created it. Likewise we must think of time as an accumulation of thought waves.

Thought waves accumulate into cycles upon countless cycles in the forming of bodies. As thought waves add density and other mass dimensions to the bodies they create, they also add time by lengthening the time intervals needed to repeat that body.

Thought waves "store up" time as they store up mass. Bodies of matter are "wound up" thought waves. The time consumed to polarize a thought wave cycle is so incredibly fast that their reproductive frequencies reach out through the universe at the rate of about 2,000 miles in one hundredth of a second. When they wind up into masses of waves to create bodies they slow down their repetitive frequencies and thus lengthen their cycles of growth and decay in proportion to the mass of thought waves which have been wound up into a formed body.

Thought waves which create a body of sound must unwind their accumulations. That also takes time. The sound of a pistol shot is a body of accumulated thought waves. These must unwind and rewind before they can reproduce a sound body. For this reason the sound can reproduce itself only 1100 feet away from its source in one second while the thought wave of its source can circle the earth seven times in one second.

The growth-decay-life-death cycle of a tree well exemplifies this principle. Fifty years of time may be consumed during one period of accumulating thought wave patterns by unfolding from its seed, and voiding them by refolding the record of those patterns back into their seed.

Life-death cycles of insect bodies vary from minutes to months. Animal life-death cycles reach into the centuries, while thought wave accumulations of solar and nebular systems reach into the hundreds of billions of years for one vibration frequency which is one life-death cycle.

Periods of gestation likewise lengthen in duration in proportion to the accumulation of the recordings of thought wave patterns upon other thought wave patterns which produce complex bodies.

All other cycles within cycles likewise vary in similar proportions; cycles such as respiration, pulse, sleep, digestion and other frequencies of repetition.

The fact of importance to know in relation to vibration frequencies is that no matter how complex the formed body may be, and no matter how great its duration in time, *the process of growth of every cycle is the same without the slightest variation. Every growing thing must pass through nine stages in this three dimensional universe of timed motion from the zero of its beginning to its zero ending.*

Every cycle is a complete octave wave—and every octave wave is a series of eight tones, the amplitude tone being two, united as one, and an inert gas—the total being nine.

Atomic structure would be difficult to comprehend in principle without comprehension of the above mentioned facts.

One must be able to vision a sun in the heavens whose duration is billions of years, and the sun which centers the tonal wave of a harp string of a hundredth of a second duration as being one in principle. The difference lies in the amount of time which must be expended in unwinding that mass of thought wave patterns into its thought wave units. Likewise one must be able to vision the interchange between the sun of a solar system and its black hole counterpart on the other side of its vacuous mate, as the same simple EFFECT of the same ONE CAUSE.

Much confusion will disappear when knowledge of all CAUSE and EFFECT is thus simplified.

Confusion regarding the many seemingly different particles of matter will disappear when one knows that each seemingly different particle is but a different stage in the growth of an elemental tone— and that each element is a stage in the growth of an octave wave cycle.

Just as a man is the same flesh, blood and bone in each of his stages of growth, so are all particles the same ultra-microscopic unit vortices of motion which are changing their pressure conditions during their whole life cycle journeys of simulating different substances.

Confusion will likewise disappear for those who search for the life principle in matter, when they know that what they assume to be life is but motion multiplying its pressures to simulate the IDEA of life, and then dividing them to simulate the IDEA of death.

XXX.
OCTAVE WAVE CYCLES

48. In order to comprehend the great simplicity which underlies the seemingly complex series of nine octaves which constitute the periodic table of the elements, together with the simplicity which underlies atomic structure, it would be well to paint a word picture of

Nature's basic desire and her simple manner of attaining her desire.

Let us therefore vision a man who is lying down to rest. He is in thorough equilibrium with his environment, for every part of his body occupies the same pressure relation with the earth's center of gravity. In this balanced position he is without the strains and tensions of electric division of his equilibrium.

This position of unchanging pressures is in a plane of ninety degrees from the radial direction of changing pressures which reach outward from the center of gravity into space.

The moment this man desires to demonstrate action for the fulfillment of his desires, he must rise from his plane of rest until he acquires that radial angle of ninety degrees to it.

Even though he can find balance when thus standing erect he must be awake and his senses alert in order to maintain balance. Otherwise he would fall to the zero level from which he rose. The reason for this is because he has divided his balance into two equal balances which are controlled by the one centered in him.

Eventually he can no longer electrically control his own balance against the resistance of the two opposing conditions he has set up by extending gravity into the forever changing pressure conditions which exist in radial directions. The polarized condition which he created by his desire for action now expresses its desire for the one balanced condition of rest and returns his body to the zero of equal pressures from which it rose.

The above is a true symbolic word picture of every action-reaction of every happening to every body in the universe.

It is also a true picture of growth-decay and life-death sequences. I shall now relate the above universal principle to the octave waves of the elements of matter and to the gyroscope principle which controls the octave periodicity of the elements of matter. I shall also describe how the gyroscopic principle cooperates with the north-south magnetic poles, which control the extension of polarizing thought wave bodies from their fulcrums to their wave amplitudes, and east-west poles, which control the withdrawals of depolarizing thought wave bodies into their fulcrums from their amplitudes.

XXXI.
INTRODUCING THE GYROSCOPE INTO THE
OCTAVE WAVE

49. The relation and purpose of the gyroscope to the wave structure of the nine octave periodic table of the elements is a very big subject for a brief treatise. For this reason I can but touch upon it lightly, but with sufficient clarity to give full comprehension of Nature's principle and process.

As all of the one hundred and twenty-one elements, isotopes and inert gases, which are produced by the electric wave machine in Nature's workshop, acquire their seemingly different properties because of the gyroscopic wheels which spin them into their various conditions, it is necessary to know how Nature causes the same kind of units of motion to appear to be so many different substances.

The present concept of atomic structure has no resemblance whatsoever to Nature's processes for there is no place within wave mechanics for it to fit into. This universe consists solely of waves of motion. Any theory which cannot find a fitting place within the wave has no other place for it in Nature.

The present day concept of atomic structure is based upon concentric shells, one within the other, which become the basis for revolving electrons placed according to formula upon those shell strata.

Centering these geocentrically and geometrically placed electrons are nuclear groups of separately and oppositely charged protons and photons. By adding one electron to an outer shell an element next in number is produced. Conversely, it is believed that if one electron could be knocked out of an element, such as mercury, the next succeeding element—gold—could be produced.

Insofar as Natural Law is concerned one might as well say that if one of the children of a French family dies it would change the family's nationality to Italian.

Transmutation will be impossible until science realizes that atomic structure is gyroscopically controlled.

XXXII.
THE NUCLEUS IS THE HUB OF THE GYROSCOPE
WHEEL

50. The nucleus of every atomic system is a single compressed

mass like the sun of our solar system. The nucleus is the highest potential and the greatest mass in its system. It is held together by the polarizing power of gravity against the resistance of the depolarizing power
of radiation.

Every nuclear mass must first be "wound up" spirally by centripetal force before it can be spirally "unwound" by centrifugal force.

Just as men and women must approach maturity before they can
bear children, so must suns be near their maturing points before they
can bear planets to become atomic or solar systems.

51. Centripetal force is generative. It polarizes bodies from their
source to their maturity. Centrifugal force is radiative. It depolarizes
bodies and voids them at their source where motion ceases.

*Centripetal force is the condition of gravitation which compresses
thought waves into body forms.*

*Centrifugal force is the condition of radiation which expands
thought waves to void form.*

52. If all people would impress this fact indelibly upon their
Consciousness it would clarify all the mysteries which beset humanity.
To really be aware of this fact is to be aware of the REALITY from
which a simulation of reality extends in forms of Mind imaginings and
returns to rest from the electric strains of Mind imaginings.

All things live and die—grow and decay—breathe in and out—
cool and heat—compress and expand—solidify and liquefy—awaken
and sleep to the mighty rhythm of the electric pendulum of the cosmos.

53. Growth-decay of the elements is the same process as growth-
decay of a tree—or of the life-death cycle of a man. The childhood-
boyhood—and youth cycles of a maturing man are the same effects
as the lithium--beryllium and boron cycles which precede carbon in
the elements.

These first, second and third elements of the octave are considered
to be different substances, each having different conductivity, density,
malleability, tensile strength, potential and melting points. Science
has not thought of these as being earlier stages in the growth of carbon,
as one thinks of the growth of a man. However, science must begin to
think that way in order to comprehend the simplicity of transmutation.
A child, boy and youth are the same flesh and blood of his mature manhood. His appearance in each of these stages is utterly different. Likewise his attributes differ utterly in each cycle.

In his childhood cycle he desires such toys as rattles and dolls. In his boyhood cycle he utterly discards these for toy soldiers, bicycles, cowboy outfits and juvenile books. Later he discards these as his desires change for higher studies, football, skiing, golf and preparation for a career.

This process of growth is universal. As we see it in tree, violet, man, elephant or insect we must likewise see it in the elements of accumulated time, or in the incredible speed of basic time. Every creating thing is based upon the wave, and the wave is a growth from a point of rest to a point of rest through gravitation, then back to that point of rest through radiation.

We must learn to think of all accumulating matter as retarded time which lengthens its intervals in the ratio in which matter appears.

Likewise we must think of time as a rhythmic illusion of motion sequences. Time appears only when motion in matter begins. Time disappears when motion sequences end.

Time is but the recorder of change. Remove change and time is likewise removed.

If one lived in perpetual light—or in perpetual dark he would be living in a timeless universe. He could then create the illusion of time only as Nature creates it, by counting the sequences of his breathings, or his sleepings and awakenings—or his hungerings.

Time sequences are the wave reversals which swing Nature's pendulum between the births and deaths of all appearing-disappearing things.

Life is but a reversal of death—and death likewise is but a reversal of life.

Time counts births and adds them up into years, and centuries, and milleniums—but time also subtracts deaths from births to remain the zero which time is.

For time lives with life and dies with death as you and I and all things else likewise live and die to forever live again in this eternal universe of eternally repeated illusion.

The senses record only the forward flow of time—but there is a backward flow of time which voids time, as there is a backward flow of life toward death which voids life.

This is a zero universe of EFFECTS which seem—but are not.

The fulcrum of the universe from which actions and reactions extend and return might be likened unto a mirror. As the action walks away from that mirror it also extends the mirror's image, which walks away with it. The reaction simultaneously walks the other way with its mirrored fulcrum ever centering it, to compensate and void the action.

Both action and reaction then come to rest and simultaneously withdraw within their fulcrum to regain the needed vitality to repeat the action of gravity and its radiative reaction.

53. Every effect of motion is voided as it occurs, is recorded (in its inert gases) as it is voided, and repeated as it is recorded.

Time has no existence. Entries in The Book of Time are but the mathematics of polarity reversals. As Nature adds up reversals of polarity she also adds their rhythms into deeper tones of less vibration frequencies of retarded time. When Nature subtracts reversals she also multiplies the rhythms of electric wave vibrations which pulse in unison with those rhythms.

That is the sole reason why man senses time. *TIME is but the pendulum of motion.* Its office is to record the heartbeat of two-way motion. *Without reversals of motion TIME has no existence.*

Time is but one of those many illusions which deceive man *into believing that unreality is reality.* It even deceives the greater savants of science into their attempts to design the shape of this shapeless universe. All of them include time in their imagined shapes of the universe.

Every point in the universe is an infinite mirrored extension from every point. Each point is the center of universal extension into that mirrored infinity which ends at its point of beginning. The universe therefore, can have no shape.

By thus removing time as a reality in Nature, and by learning to think cyclically, in the orderly rhythmic simplicity which Nature applies to all creating things, one will be greatly aided in his endeavor to see the universe as one whole.

With full understanding of this pulsing heartbeat principle of interchange between the two opposites of electric expression, the expanding universe theory would never have been conceived.

XXXIII.
ALL SYSTEMS ARE EXPANDING SYSTEMS

54. Just as a business or family, or any organized group must first

generate a nucleus for the expansion of an idea into a system, so does the universe generate a nucleus for extending its idea into systems.

Naturally the universe is expanding and extending its generated nuclear masses into systems but it is forever contracting masses into nuclei in order to extend them into systems.

Nature first generates matter by polarizing it into a spherical nuclear mass. She then radiates matter by depolarizing it into expanding systems. Every system, whether atomic, solar or nebular, is expanding in relation to every other system in the universe, and is also expanding as of itself.

55. Planets are born from rings thrown off from the equators of suns. Moons are born from rings thrown off from planets. Rings "wind up" into planets and moons. These contract as they wind until they become spheres. They then expand as they "unwind" into oblating spheres.

All suns and moons of stellar systems are created only by compressing electric waves. The theory of the accumulation of dust clouds into matter is not true to Natural Law. There is no such dust in space, not even the weight of a milligram.

Space is an equilibrium which is polarized into four octave waves of invisible matter, but octave waves of matter even though invisible, are not "dust." These space octaves will be referred to later.

All planets and moons of their systems spiral farther and farther away from their primaries. They also spiral outward from their own axies of rotation, which shorten as their equators lengthen. They gradually "swell up" into many times their original size as they expand.

When Jupiter was where the earth is it was not more than twice the size of the earth. It is now very many times larger. It has likewise expanded by throwing off rings such as you see around Saturn. These have become moons. Four of them are still on the plane of Jupiter's equator. It is even now preparing to throw off more rings which are seen as belts circling its equator.

POSTULATE

56. Centripetal spirals multiply gravity to form the nuclei of borning systems, and centrifugal spirals divide gravity to radiate nuclei into dissipating systems.

* * * * *

Matter contracts into solids by "squeezing" space out of it. This draws its particles closer together and decreases its volume. Matter expands by "swallowing" space within it. This thrusts each particle farther apart.

57. *The north-south poles control generation into form. The east-west poles control degeneration of form back to its source. When true spheres have been generated the four poles unite as one and then reverse their directions. Prolation then ceases and oblation begins.*

This universe of EFFECT is dual. It is a divided universe in which each positive half of every effect is balanced by its negative opposite half. For each hot sun there is an equal cold vacuity awaiting in space to born another sun. Tides when lowering are simultaneously rising, and day cannot come without balancing night.

Our senses can detect the expansion of matured masses into systems, for the fiery arms of expanding systems are visible in thousands of nebulae. Hot radiating masses are visible but cold generating spirals which are creating hot bodies are not visible. The two black arms in every nebula are generating and contracting the hot radiating bodies which arc its fiery arms. See figures 131, 132 and 133.

POSTULATE

58. *Wherever there is motion there are two magnetic poles to control their contraction into "matter," and another two to control their expansion into "space."*

When matter swallows space matter disappears. When space swallows matter matter reappears.

XXXIV.
OBLATING SPHERES

59. It must be known that the north-south polarity which divides the universal condition of rest into two opposite conditions of motion to create matter must have a counterbalancing polarity whose office it is to void the two opposed conditions of motion to restore the condition of rest.

Each of these oppose each other. One pair gains ascendancy for one half of a cycle. The other pair then gains it. This principle is demonstrated by the life half of a life-death cycle being stronger than the death half—then the death half becomes stronger until the cycle is completed.

In either half of the cycle polarity controls its balance, but the office of north-south polarity is to prolate mass from its beginning at the base of a cone, to a sphere at the cone apex by extending its poles, while the office of the counterbalancing polarity is to oblate mass from a sphere to the base of a cone, by extending its equators.

In an oblating sphere like our dying planet, the east-west polarity has gained the ascendancy. These two poles control and balance the extension of the earth's equator, the expansion of its volume and its orbit into ever lengthening ellipses as the earth gradually flattens and increases its distance from the sun.

60. The converse of this effect is exemplified in our prolating sun. It has not quite matured into a true sphere. Its north-south polarity is still preponderant and will continue to predominate until the sun reaches true sphere maturity at its half cycle point.

Aeons will pass before the four poles unite and reverse their positions and directions, which will begin the flattening of the sun at its poles and its eventual disappearance by throwing off sequential giant rings.

The sun is still prolating while its planets are becoming increasingly oblate. The moment that earths or moons begin to oblate, that moment their equators leave the plane of the sun's equator and their elliptical orbits are extended by the extension of their two east-west foci.

Newly born planets and moons, like Mercury and the four inner moons of Jupiter, hold to their planes of birth on the sun's equator until they begin to flatten.

XXXV.
UNBALANCED ATOMIC, SOLAR AND STELLAR SYSTEMS WOBBLE

61. When a top spins swiftly enough upon its axis to maintain an angle of 90 degrees from the ground it spins without wobbling on its axis, for it is in balance with gravity. Its axis points directly toward the earth's center.

When spinning slows down its center of gravity is divided. The top then wobbles. We say that it is out of balance. Scientifically expressed we should say that its balance is divided.

To divide balance its one center of gravity must be extended to two foci instead of one. We exemplify this effect.

62. Two children play seesaw by alternately lengthening opposite ends of their lever. That divides gravity by throwing the lever off center from its fulcrum. The fulcrum seemingly moves toward the short end of the lever to counteract balance.

When the reversed motion takes place the fulcrum seemingly moves to the other side of its own center to again counteract balance. That develops two seeming extensions of the fulcrum from its own center. The fulcrum has not moved however, for the fulcrum is gravity. It has but seemed to move to two east-west points. It has seemed to make gravity oscillate between two extended balance points.

63. These two east-west extensions of gravity are east-west magnetic poles for they extend as such only for the purpose of keeping this universe in balance in its every effort. When the seesaw returns to its level these two east-west foci withdraw into their fulcrum and cease to be—because unbalance has ceased to be.

When the spinning top slows down it leans away from its vertical axis. It has become unbalanced with its north-south vertical axis which points directly toward the center of the earth.

That leaning describes a circle around its perpendicular axis. North-south is seemingly divided into an extended pair which seems to cause gravity to oscillate.

64. *Those seeming oscillations are east-west extensions of balance which counteract and control any unbalance which threaten to upest the balance of north-south extensions.*

65. The above paragraph is frought with meaning which must be clarified. The relation of balance to gravity is so little understood that it should be made clear.

We see a man on a tightrope extending a balance pole east and west from his south-north direction to counterbalance any unbalance he may create.

We relate that fact to gravity in a far too vague way of thinking. It should not be vague. We should KNOW its meaning dynamically.

66. Its explanation is given in two steps. The first step is to reduce it in principle to utter simplicity. The second step will be to amplify that simple fundamental.

FIRST STEP

67. We must first realize that Creation is but the electric thinking of idea expressed by moving body forms imagined in the Mind of the Creator. The moving body forms are created in the image of the Creator's imaginings. *The body forms are not idea, they but simulate idea.*

When man creates thought forms for his ideas his conception expands from the zero point of its beginning. He builds a complete mental three dimensional form for his idea and creates a body to simulate that idea. He then fatigues from thinking that idea and rests for an interval before again thinking it into further form. The thoughts which he extends to cause body forms to appear he now retracts and they disappear.

Body forms of matter appear when Mind concentrates by thinking idea. They disappear when Mind decentrates to rest from thinking idea.

Mind of God and Mind of man are one. The Creator thinks idea as man does—by extending it in waves of electric thinking and withdrawing it by reversal of extended thinking.

Briefly then, we may define Creation as a Mind imagined electric extension from a point and its retraction to that point.

SECOND STEP

68. *This universe is the sum total of electric actions and reactions expressed in thought waves of two-way motion.*

Every action is an outward radial extension of balance from one balanced condition to create two opposite equally balanced conditions.

Every action is therefore, an outward radial extension of balance from a centering point of universal balance.

69. The extension of infinite radii from a dimensionless point brings into being a three dimensional radial universe. It has length, breadth and thickness. And it has form, the *sphere.*

Also it has measure—the measure of the energy which desire for extension gave to it. The desire to divide and extend one condition of rest into two intervals of motion are marked throughout the universe by the magnetic surveyor and controller of balance.

70. North and south poles are measured out to limit the extension of form from its wave axis, to a sphere of balanced curvature.

East-west polarity resists north-south extension of matter beyond the form of a sphere. Its office is to return the sphere to its wave axis. North-south polarity resists that change.

71. In a radial universe of varying pressures the change in pressures is only in the inward-outward direction. There are therefore, only two directions of changing pressures in this universe.

72. The inward one is the direction of multiplied pressures. Gravity means multiplied pressure. Gravity is north. North is positive.

73. The outward one is the direction of divided pressures. Radiation means divided pressure. Radiation is south. South is negative.

74. North-south is the direction of dynamic action. The piston of the universal heartbeat is north-south. Cyclones, tornadoes, lightning, rains, and all other dynamic effects of motion are north-south. Their potentials all multiply in the direction of north—and divide in the direction of south. Solidity of matter is north. Emptiness of space is south.

75. North-south represents the divided universe into pairs of equal opposite conditions—the condition which we call *gravity*—and the condition which we call *radiation*. Briefly stated, north-south is the direction of motion and time, for sequences are north-south reversals which born time.

SUMMARY

76. East-west poles are measured intervals of extensions on planes of rest. They represent the *undivided* universe of *unchanging* balance and potential condition.

77. North-south poles are measured intervals of extensions on planes of motion. They represent the *divided* universe of *changing* balance and potential condition.

78. North-south polarity divides the universe into two equal and opposite conditions by extending balance dually and dividing it into pairs.

79. East-west polarity resists that division and sets up two counterbalancing east-west poles to control the balance of the two conditions on their return to the one condition of rest.

EXAMPLES

80. (1) A true sphere sun is in perfect balance. It has but one focal

center of gravity because its radii are of equal length. The moment the sphere oblates its radii are of unequal length. Their potentials vary because the equatorial extensions of mass out-balance the polar extensions. Two east-west balancing poles then extend from the center of gravity to control the unbalance of potential now set up in the sphere.

A section of the sun, cut through upon the plane of its equator, would be circular. The radii of a circle are equal. A section cut through the poles would be elliptical. The radii of an ellipse are unequal.

81. A sphere has but one focal center but an ellipse has two.

During the prolation of elliptical spheroids to spherical form the two east-west foci draw closer to the center of gravity as northwest foci extend away from that center. As spheres oblate to elliptical spheroids the east-west poles extend away from the center of gravity as depolarization draws the north-south poles closer together.

82. (2) The sun is practically a true sphere. Its equatorial ring of hundreds of millions of miles is a gyroscopic wheel. Its shape is circular at its equator but its pressure directions are spiral. The planet Mercury is practically a true sphere. It is an equatorial extension of the sun. If it were still a part of the sun's body it would revolve around the sun's axis as an integral part of the sun's body.

83. Even though it has separated from it and has an axis of its own upon which it must turn, it must still revolve around the sun's body as well as its own.

84. Mercury is also a gyroscopic wheel. Its ring extension coincides with the ring extension of the sun. They are on the same plane, therefore their poles of rotation are parallel. If the pressures of the ring were equipotential circles, Mercury would describe a circular orbit around the sun, but they are not—they are spiral, therefore there are perihelion and aphelion foci which balance and control the extension and retraction of its orbit around the sun.

This same thing is true of the four inner moons of Jupiter and the inner moon of Mars.

85. Our earth is not located upon the plane of the sun's gyroscopic ring. It has broken away from it to an angle of 23 degrees. Its gyroscopic disc is so greatly out of balance with that of the sun that it has to revolve around the sun below the sun's gyroscopic disc for one half of the year and above it for the other half, instead of keeping on the same plane with it as Mercury does.

That puts the earth in the same predicament that the tightrope walker finds himself when he leans out of balance with gravity, or a spinning top is when it leans off center.

86. The angle of their leanings is the same in principle as the leaning of the earth's axis. It makes the earth wobble on its axis to describe circles around its plane of gravity, while it reaches out for two counter-balancing foci just as the tightrope walker reaches for two counter-balancing foci.

87. Science has been retarded in discovering this fact of counter-balance of polarity by misinterpreting the action of the tightrope walker, the interpretation being that the extension of a balancing pole is to counteract the weight of his unbalance by extending an equal weight upon the other side of his balancing equator.

88. *That is true but the conception of weight is not true.* By reading my chapter on weight this will be clarified by the true conception of weight as the potential of resistance to strains and stresses set up by any departure from a balanced condition.

89. When the earth "stood up straight" it had no need of extending its balance pole, but the moment it leaned it needed those counter balancing foci as much as the tightrope walker needed them.

XXXVI.
WOBBLING GYROSCOPES SEEK BALANCE

90. *Mass is motion and motion must be balanced by opposed pairs of poles. When motion ceases polarity likewise ceases.*

Motion does not cease however, until extended mass returns to the wave axis from which it was projected. The moment it again leaves that axis in opposite two-way extensions, poles appear because balance is divided and must be controlled.

91. Tops spin on their pegs and solar and atomic gyroscopes spin on their hub shafts, but the principle of their wobbling is the same. They wobble when their shafts are off center.

The hubs of gyroscopic wheels do not center their rims in the first three pairs of tones of the octave. The wheels are ellipses and the hub of the wave shaft is gravity, so gravity does not center the wheel for the first three octaves.

A metal gyroscopic wheel, or flywheel, multiplies centrifugal force as it increases in speed, but Nature's atomic gyroscopic wheels are centripetal vortices which contract around their shafts. They are like whirlpools, or cyclones which pull inwards and multiply centripetal force as they thus contract to form hubs for their wheels which are centering suns.

92. Two children cannot move while they are in balance with their fulcrum, for motion is impossible in an equilibrium. Balance must be divided into unbalanced opposite pairs before motion becomes possible.

Nature likewise cannot produce motion without thus dividing balance to produce two opposing conditions. Centripetal force thus produces carbon when its speed has multiplied sufficiently in each succeeding tonal effort to find a balance between those two opposing conditions.

Wobbling gradually decreases as the prolating spheroidal hub of the gyroscopic wheel contracts to a true sphere and the shaft of gravity centers the hub, and its north-south poles are parallel with the wave axis of its beginning.

XXXVII.
HOW GRAVITATION AND RADIATION
BORN EACH OTHER

93. Nature works in strange ways. Of all her mystifying processes her manner of producing the double polarity which assures two-way balance for the two-way journey of her two conditions, is perhaps the most illusive of her illusions. It is well to clarify this mystery step by step at this point.

(A) The carbon wheel spins true upon a *horizontal* shaft which arose *vertically* from its plane of equilibrium.

(B) The rim of the wheel begins to spin on the *horizontal* plane of equilibrium and arises to become the *vertical* equator of its hub.

(C) The vertical has become horizontal and the horizontal has become vertical to transform one unchanging rest condition to two changing conditions of motion.

(D) The positive electric worker has made the *rim* of the wheel become its *hub* by use of its centripetal force. That is how Nature manufactures GRAVITY and multiplies potential to contract waves into solids surrounded by space.

(E) The negative electric worker has made the *hub* of the wheel become its *rim* by use of its centrifugal force. That is how Nature manufactures RADIATION and divides potential to expand waves into space *centered* by solids.

(F) The rim of the wheel is now 90 degrees from the equilibrium plane of its birth, and is 90 degrees from the shaft of its hub. From a plane of no motion it has become a sphere of maximum motion.

(G) The hub shaft of the wheel is now parallel with the plane of rest and 90 degrees from the plane of maximum motion.

(H) The rim of the wheel was maximum speed and the hub was minimum when motion began on the plane of rest, but now the hub is maximum speed and rim is minimum when the wheel stands up from rest.

94. This is nature's process of dividing the still Light of the Creator into the two moving lights of matter and space to simulate the Mind imaginings of the Creator by moving image forms of His Creation.

A word picture of this process might simplify Nature's method. Imagine therefore, the seed of idea placed upon the wave axis like the seed of a tree put into the ground.

Now imagine the ground rising as a hoop would rise from the ground until it stood straight up instead of lying down.

As the ground rises to stand up, imagine the idea of the tree unfolding in a series of four efforts which we will call stages of growth.

The seed of the idea becomes a fully formed mature body when the ground has arisen from wave axis level to wave amplitude height, 90 degrees from its axis level.

The ground, which borned the formless seed, is now the vertical equator which balances the fully formed body.

Half of the upright tree extends to the east of that equator and the other half of the west of it. Its roots extend north toward gravity, and its branches radiate south toward space.

This is the manner in which growing matter appears.

Now comes the reverse process. That fully formed body which has unfolded from its seed must now refold into its seed. This it does in four reverse stages of decadence, and as it thus refolds the ground lies down gradually with all of the body still contained in it but refolded as patterned seed.

This is the manner in which decaying matter disappears.

This visualization pictures that method of Nature which borns and reborns its patterned ideas forever and forever without end. That which comes from the ground must return to it for rebirth. Patterned forms must disappear into their seed and be added to at each rebirth.

Idea is eternal. Bodies which manifest idea are transient but their repetitions are eternal. There is no exception to this process of repetition of bodies which is called reincarnation when applied to man. The process is universal however, and applies to all creating things—not man alone.

95. If one would know the heartbeat of the universe one could know it by comprehending this rhythmic balanced interchange between the pairs of opposite conditions which gave eternity to this universe through eternal repetitions of living-dying sequences.

Thus it is that the life-death-growth-decay process of division of an equilibrium into two oppositely conditioned states of motion is repeated in every action-reaction of motion, no matter how simple, or how great.

POSTULATE

96. *All matter begins its accumulation from cone bases on wave axies. It multiplies its accumulation while spiralling to cone apices. It ceases to accumulate when it becomes a sphere—and redistributes its accumulation in rings around the sphere to become cone bases on wave axies for repeating its accumulation.* See figure 131.

Verily the sphere—which carbon is—has arisen from its resting place to stand up and go into action in a divided three dimensional universe of change for just a little while, before lying down to rest in the undivided universe of stillness in order to regenerate vitality for again arising into action.

XXXVIII.
THE NINE OCTAVE PERIODIC TABLE
OF THE ELEMENTS

97. The periodic table of today lists 92 elements, including isotopes and inert gases. Many listed as elements are isotopes, which are divided fractional elements.

My periodic table lists 63 elements, 49 isotopes and 9 inert gases making a total of 121.

From Nature's point of view there is but one element—CARBON and but one form—THE CUBE-SPHERE.

Carbon crystalizes in the form of its wave field, which is a true cube. The nucleus of its system is a true sphere. The plane of its system is 90 degrees from its wave axis, 90 degrees from its pole of rotation and 90 degrees from the axis of its north-south poles. The shape of the carbon atomic system is a disc, as shown in figure 131. The orbit of every planet of the carbon system is on the plane of the carbon equator, and that equator is on the plane of the wave amplitude.

Carbon thus manifests balanced form in body and unity in balanced sex mating. It has but one equator. All elements which are not on wave amplitudes are disunited pairs which are divided by three equators. Each single element is divided in itself by its own equator and each pair is divided by the wave amplitude equator.

98. *Carbon symbolizes the marriage idea in Nature. Its one equator is the bond of its unity.* It is no longer a pair—and that is what marriage in Nature means, and what it should mean in man's mating practices. Divided pairs have opposed attributes. The negatives of pairs are metallic acids—the positives are metallic alkalis. *All are conductive, for conductivity is a search for balance.*

Balanced unity voids acidity, alkalinity, metallic quality and conductivity. By eliminating these qualities carbon becomes a salt—which means a mineral with the qualities of stone.

When disunited equal and opposite pairs "marry," such as sodium and chlorine, they likewise have but one equator instead of three the instant they unite as sodium chloride. They likewise lose their metallic, acid, alkaline and conductive qualities and crystallize as true cubes.

An example of unbalanced mating in Nature is that of the marriage of sodium and iodine or sodium and bromine. Each of these marriages have stability but there is a residue of unbalance in each of them which is evidenced in distorted cube crystals. Each of them would likewise continue as harmonious marriages unless chlorine appeared, in which case Nature would immediately annul the marriage in favor of chlorine.

99. Carbon has the highest melting point and greatest density of all the elements. This means that carbon is also the most enduring of all elements because of having accumulated more time cycles. It likewise means that carbon is the least radioactive of all elements because radioactivity only begins to express itself by outward explosion at wave

amplitude, although it is strongest at that reversal point where genero-activity and radioactivity meet.

It is from Nature's point of view we will very briefly describe the nine octave cycle of the elements, with the hope of unifying man's point of view with that of Nature.

100. The one supreme outstanding characteristic of this electric universe of two-way balanced effects of motion is the cyclic unfolding of matured body forms to manifest MIND IDEA, and their refolding into the Source of all IDEA.

Bodies of matured forms are unfolded by a series of four efforts in positive-negative pairs. Likewise they are refolded by a reverse series of four efforts in similarly mated pairs.

101. Each effort in Nature to unfold and refold is a stage of in-ward-outward growth toward the formation of a matured polarized body, and away from it toward its seed idea.

The fourth positive-negative pair of every octave is united as one. See figures 87 and 114. They unite as one at their wave amplitude, which in every wave points directly toward the center of gravity. These two united efforts constitute the matured body form of conceived idea.

They are the meeting points of life and death—the reversal points of rest which divide generation and radiation. At that meeting point is the greatest density, highest melting point and highest potential of the entire cycle.

In that united pair is the matured body of the one element— CARBON.

Every completed idea in Nature is expressed in nine efforts—or stages—which are eight octave waves, plus the matured centering amplitude wave of the whole nine octave cycle.

102. Each octave of the elements grows from its inert gas just as a tree grows from its seed. The inert gases record and store for repetition all that has gone before in that octave.

103. In the Mendeleef table of the elements hydrogen is shown without an inert gas. This is as impossible as producing a child without parents.

Hydrogen is also shown as being the only element in a whole octave. That is also as impossible as charging only one of the two cells of a battery.

104. *Hydrogen is not one element, but eight. It is a whole octave*

in itself but Nature has not made it possible for the senses of man to detect this easily.

When I explained this fact many years ago to science it went into research and found other tones of this octave which it mistakenly called isotopes. What science found were full tones, not isotopes. Science had numbered the elements from 1 to 92 however, on the presumption that there were no others, and had no alternative but to call them isotopes.

105. In the Mendeleef table series 5-7-10-11 and 12 are shown without inert gases and without being full octaves. These series are also partially filled with isotopes which do not belong in the groups in which they are placed. Also a group numbered 8 consists of nine iso-topes to which full numbers have been given. In fact all isotopes are numbered as though they were full tones.

106. Isotopes do not appear in Nature until the 6th octave, and then only between 3 and 4 positive and 4 and 3 negative. They in-crease in numbers in the succeeding older octaves because the aging carbon is unable to reach the true sphere in either of them. Its many at-tempts to do so result in producing many isotopes.

Like the fully matured strong man who keeps his vitality for a long period of time, carbon rises again to amplitude at silicon as a non-metal, but from there on the gradual radioactive decline makes it im-possible for another balanced nonmetal to appear at wave amplitude.

107. The fifth octave is the balancing one of the nine which Nature demands in all of her expressions. That is the octave of matured vitality. The four older octaves are fully evident to our senses because they have accumulated density by accumulating time cycles.

The four younger octaves are beyond our sense range with the ex-ception of hydrogen, which has been listed as only one of that octave.

These exist in Nature for Nature is balanced. It must have the four younger octaves to counterbalance the older ones.

As I have heretofore said, one can know many things which he cannot sense. One can therefore, KNOW that balance in Nature's pol-arization principle DEMANDS equality of division in all of her paired effects.

It is not just necessary to KNOW this fact however, to be con-vinced of its truth, for it can be proven by reading the history of the elements from their beginning in spectrum lines. The red lines in the spectrum of hydrogen do not belong to one octave alone. Each red line tells of another invisible octave. Spectrum lines should be read as accumulated time in history, not as though all the lines of any reading belonged to one element of one octave.

108. The reason for the intervals between these red lines in the spectrum is not because they represent the pressures of one element but because each sequential octave increases in density, which also retards time sequences.

The reverse of this principle applies in depolarizing bodies. Depolarizing bodies on the radioactive half of any cycle project time accumulations from them at tremendous speeds. Helium and other inert gases explode outwardly from tungsten at approximately half the "speed of light" while similar "rays" explode outwardly from radium, actinium, thorium, uranium and uridium at almost the speed of light.

Conversely, generoactive rays explode inwardly at tremendous speeds in the first three invisible octaves. Alpha, beta, gamma and "cosmic" rays explode inwardly to center invisible generating matter as they and the older inert gases explode outwardly from degenerating visible matter.

The nine inert gases which form the seed patterns of unfolding matter mystify observers who do not comprehend their action or their purpose. The refusal of inert gases to combine with elements has always been an insoluble mystery.

After scandium in the 6th octave and arsenic in the 7th octave, five separate efforts are needed to produce cobalt. Carbon is still tremendously strong of body in its cobalt stage but cobalt is not a true sphere, nor is its wave field a true cube. For this reason cobalt is metallic, and so are the carbon prototypes in the rhodium and lutecium octaves.

Naturally such isotopes as cerium, thorium, tungsten and many others, also show their direct relationship to hydrogen in many ways, such as inflammability.

Carbon itself gives much evidence of its identity with hydrogen. Every chemist knows that carbon is the basis of all organic and inorganic matter, and that hydrocarbon compounds are more numerous in Nature than any other combinations.

Flesh leaves a residue of carbon when acted upon by acids. Carbon is the basis of all vegetable growth as well as animal, as evidenced in the earth's coal deposits and the charcoal of burnt wood.

Likewise hydrocarbons will not react to acids or alkalis because acids and alkalis are voided in the elements when they find the perfect balance of gravity in the true cube wave field.

Carbon is the only element which completely measures up to that requirement. Hydrogen so nearly measures up to it that it is immunized from reaction by acids or alkalis when in combination with carbon.

These facts are cited in order that the metallurgist and chemist will base their thinking upon the growth-decay or life-death principle of matter rather than on the idea of many separate substances.

By dividing the entire nine octave cycle into its two opposite half cycles, one half being generoactive and the other half being equally radioactive a comprehensive base for transmutation will replace the present concept of dislodging electrons, or adding to them to transmute one into another.

The age of transmutation will come only through the transformation of man, and man's transformation can come only "by the renewing of his Mind" through new knowing. It has ever been that way since the dawn of Consciousness, and it will ever be.

Whenever new knowledge of a transforming nature permeates the race the standard of world culture rises. The art of the Italian Renaissance transformed mankind from seven centuries of Dark Ages. *New knowledge of Natural Law is slowly driving superstition out of man.*

Spiritual knowledge has transformed mankind step by step from his jungle age. Scientific revelations have also transformed man step by step since early thinkers rediscovered that the earth was round, after having forgotten it for over ten centuries.

Man thinks differently at each transformation from new knowing, whether religious, philosophical, scientific or artistic. Another kind of man emerges from new standards of thinking.

XXXIX.
INDUSTRY'S POWER CREATING PROCESSES ARE STILL PRIMITIVE

Primate man discovered the flame. He began to use it by burning large quantities of the stored up gravity of earth's resources as *a large percentage of fuel to obtain a small amount of heat.*

Later he learned how to use the heat for power, but he still used a large amount of fuel to obtain a small amount of radiation for his power.

Industry now has giant furnaces, burning vast quantities of fuel for a small amount of radiation which it can use, and a vast wastage which it cannot use.

The fuel it is using is dug from the ground with hard labor, transported with great effort and shoveled into furnaces by the sweat of man's brow. Man is beginning to use the gravity of Niagaras and flowing rivers for electric power, which he wastes in radiation instead of multiplying its gravity, as Nature multiplies it in this curved radial universe.

These vast power-wasting furnaces which seem so impressively suggestive of great progress, are but the multiplied flame of primate man. They have glorified man's primacy but they have not lifted him out of it.

There is still the needless waste of earth's resources—still the burden of it in the sweat of heavy labor—still the treadmill of it which is the root *cause* of present day mass revolt.

WHAT IS THE ANSWER?

Knowledge alone will lift the industrial world from such a state of primacy.

These vast unclean, smoky furnaces and treadmill-worker slum towns will disappear when science transforms industrial power usage by "manufacturing" gravity for power usage, the way Nature manufactures it in her spherical gravity-making machines.

Nature is curved—and it is radial. This curved radial universe of step-up and step-down transformer spheres, stores up the gravity which man is so wastefully using in the hard man-debasing way.

Users of power must realize that neither gravity, nor what science calls radiant energy, are existent forces in Nature. Both of these two expressions of force are manufactured products of Nature, and man can manufacture them as readily as Nature, for he has the same equipment to manufacture them that Nature has.

That equipment is the electric current, with its resultant dual polarity, *and the curvature of both polarities.* That is all that is needed except the fuel for the electric current.

This has been the only stumbling block to unlimited power ex-

pression. Even now the use of gravity pressures in the falling waters of the Columbia River are causing "brown outs" because of the lowering river and excessive drains of industry.

Free hydrogen would end such troubles forever. It could be so simply and easily obtainable in unlimited quantities that every man, whether farmer or blacksmith or factory owner, could make it as he needs it for heat or for power, *with patents only upon the machines but not upon the fuel.*

New knowledge of Nature's manner of multiplying both genero-activity and radioactivity will make a new civilization, for it will uplift man to the higher status needed for a new civilization.

XXXX.
THE SECRET OF MAN'S POWER

Knowledge of polarity control and the dual curvature of this radial universe of multiplying and dividing radial pressures is the secret of man's new power. Science has not used this power for industry because it has been unknown. With that knowledge science could blow this planet to pieces by multiplying the power of radioactivity through the lenses of polarity curvature, in addition to chain reaction explosive power. *The power within any mass can be used against itself just as a man can—and does—use his own great power against himself.*

Through this knowledge man could electrocute or incinerate marching armies to the last man, or destroy approaching planes or ships as far away as they could be detected by radar.

Entire nations could insulate their peoples from any enemy from without. By the time that becomes possible however, there will be no enemy from without—for the thing which makes man the enemy of other men is greed for material wealth and fear of bodily insecurity.

Both greed and fear will disappear from the face of the earth when man need no longer have to kill other men to obtain all of his material needs for personal aggrandizement or bodily security, for material abundance will not be dependent upon matter.

A new power of man will be his ability to project gravity in the shape of a high potential focused from a point to a distant focal point, instead of projecting radiation only, as he now does.

An outward explosion from dynamite, for example, is radiation. It is effective for but a limited distance from the source of the explosion.

An inward explosion is gravitative, and is effective wherever projected. Its tremendous power could melt the stone of a mountain for needed metals or destroy an enemy during the interval of time needed to teach mankind the futility of enmity.

That is what I mean by the transformation of man through new knowledge. *New conditions arise from new knowledge, and man must conform to new conditions. He cannot help doing so.*

Man's nature is essentially good. The evil in man springs from fear for the safety and security of his body, and from greed for the satisfaction of bodily desires. Remove these and man will naturally respond to the good in him, for all men seek the peace, happiness and security which only a balanced system of human relations will give to him.

XXXXI.
NEW POWER FOR SCIENCE

Man's transformation by science will take much time but it can begin NOW. A beginning is a reversal of direction. To reverse the direction of the downward plunge is to begin to climb into the heights.

The first step for science is to insulate its countries from attack by other countries and thus save the life blood of its Nations and return destroying armies to useful pursuits. As very little time is needed to bring this about, after the principles involved are thoroughly understood, the threat and fear of war should pass from the mind of man forever. *Even if war should start before this had been accomplished it could not go far before it could be remedied.*

The second step should be to give the world a new and inexhaustable fuel. Free hydrogen is the logical supply because free hydrogen is the basis of the four space octaves. The entire population of ten planets like ours, could not lessen its total because Nature balances the withdrawals of gases with replacements continuously. Nature's replacements for withdrawals of solids consumes the amount of time taken to grow them.

Nature may take a million years to grow forests into coal. Coal is multiplied nitrogen, for nitrogen is a gas of carbon. *Nitrogen can be transmuted continuously from the atmosphere in unlimited quantities forever.*

The atmosphere is composed of nitrogen and oxygen. Oxygen is carbon twice removed, just as nitrogen is carbon once removed. Likewise hydrogen is carbon one octave lower, but not removed tonally. Gyroscopically, carbon and hydrogen are the same, for their planes of structure are identical.

Hydrogen could therefore, be transmuted from the atmosphere in unlimited quantities by merely changing the gyroscopic plane of nitrogen to the 90 degree angle of wave amplitude, upon which hydrogen rotates.

It would simplify scientific thinking if science would view the universe of "matter" and "space" as gravity which accumulates generoactive predominance into hydrogen in the first three and a half invisible octaves of matter, that man today thinks of as space.

The visible universe begins at the middle of the fourth octave and continues to carbon—its generoactive maximum—where fourth and fifth octaves meet at wave amplitude.

From there on radioactivity begins its depolarizing process but the "bodies of the octaves grow bigger" and keep within the visible range while dying, just as a tree, or man grows bigger of body during declining years.

If therefore, science would form the habit of thinking of matter and space in terms of the carbon octaves and the hydrogen octaves it would simplify their work of transmutation mightily.

Science should also form the mental picture of the visible carbon octaves as but a pea sized volume of solid matter suspended in the center of a great auditorium of rare gaseous matter, millions of times greater in volume. Then realize that the very small globule of many solid elements of the carbon octaves, are wound up from that vast volume of the hydrogen octaves of space.

Matter thus wound up is sequentially unwound into gases of the hydrogen space octaves, and its action-reactions are recorded in the inert gases which born each octave.

Thus matter gyroscopically emerges from space and is "swallowed up" by gyroscopic unwindings into the "space" which borned it, as has been rightly conceived.

"Space" is not empty—nor is it an "ether." The space which surrounds every particle of matter in every wave field is the negative half of the wave field. The solid nucleus is the positive half. Both halves are equal in potential but vastly unequal in volume.

The next step in habit-forming thinking is to think of matter as being the accumulation of the same thing—wave motion—rolled up in tine-layers like a snowball—the final layer being called carbon but all being different conditions and pressures of the same thing.

Add to this thought that the universe consists of wave fields within wave fields—stellar, solar and atomic in measure—but of a like "substance" and of a like structural formation. Nature has no separate method or process of creating systems. The heavens clearly evidence the unwinding of mass by the way of rings and systems, but the *senses* do not so clearly record the winding of mass as a basis for systems.

During this whole process each succeeding element becomes another phase of the same thing throughout the whole journey. *The change of attribute is due solely to the different relations of pressures and that is determined by polar relations.*

Nature does not transmute one element into another. She merely makes her progressive change of elements by a continual readjustment of her gyroscope.

Elements are tonal. One wire of a piano can become a whole octave by changing its pressure relations sufficiently to either multiply or divide its vibration frequencies. Everyone is familiar with the fact that placing a book on top of an organ pipe lifts its tone just one octave higher.

Such effects are not transmutation. They are merely changed dimensions of states of motion. All of the notes which the organist plays are but one tone multiplied or divided in rhythmic pressure relations.

That is the way the chemist of tomorrow should think of the elements, and not think of them as different chemical substances with different attributes. *Chemistry should be based upon the idea of gyroscopically changing the north-south-east-west polarities of one tone to increase—or decrease—its time frequencies.* The piano tuner uses an instrument to wind up his pressures from lower to higher tones. The chemist should use the electric current and solenoids as his tuning instrument.

The very thought structure of tomorrow's chemist should very radically change in many other respects too numerous to describe. *One of these is to eliminate from his thinking the idea of one thing becoming another. That is not Nature's way.*

In Nature one tone ceases to be and another becomes. In other words, one formula for a patterned wave vibration ceases when another measured wave vibration begins. We must also carry this thought farther by not thinking of cessations, beginnings and endings. We must think of them as awakened continuities which we can "put to sleep" when we have no further need of them, or "awaken" when we have need of them.

The electric current of the universe is ready to motivate any tone as we desire to awaken it, just as the electric current of the organ is ready to awaken any tone when the organist desires to awaken it.

We should not think of sodium and chlorine as having *become* sodium chloride—or that sound has *become* silence—for each of them *are* and always will be. We should think of each of them as another note played on the universal organ. We change its tuning pattern if we want new isotopes which Nature has not yet given us—or we unite two unbalanced halves to secure stability—or produce explosives by multiplying unbalance.

That is Nature's way. Carbon unwinds to nitrogen because of the predominant power of east-west negative polarity. Likewise nitrogen unwinds to oxygen. That does not mean that carbon has ceased to be, or that it has *become* nitrogen and oxygen. It means that carbon still *is* but it has changed its pressure dimensions, just as John Jones is the same John Jones that he was ten years ago.

Nature demonstrates this fact by "transmuting" nitrogen and oxygen back again into carbon. Every root of all vegetable growth rewinds both of them upward again into carbon. Likewise the bodies of all animals rewind oxygen and nitrogen into the proteins of their flesh, bones, horns and hair.

The roots acquire the complex formula for rewinding into violets, pine, oak or apple trees—or of man or bird—from the inert gases of their octaves which have recorded the unfoldings of the many ideas of Nature in the seed of these ideas.

As Nature unfolds from the seed to record its patterns in moving body forms it simultaneously refolds into its seed in order that the refoldings can be repeated in like patterns.

All of Nature's many forms are patterned "motion picture" projections from the still seed patterns stored in the inert gases, to make the "positives" of body forms. The reverse direction of retraction creates the "negatives" of those body forms. The principle of photography is applied throughout Nature.

All unfolding and refolding patterns are gyroscopically manipulated, electrically motivated and magnetically measured and controlled.

The above stated facts makes it necessary for the chemist of tomorrow to make use of the electric current, the solenoid and systems of polarity measurements of gyroscopic planes, to do in his laboratory what Nature does in its laboratory.

Nature "puts a book" upon the top of her organ pipe of the nitrogen tone to produce its octave harmonic—phosphorus—and again to produce the next octave tone above—which is arsenic. Nature does likewise with oxygen to produce sulphur and selenium.

Today's chemist makes wasteful and complex use of the electric current, crucibles and other equipment. *The electric power is wasted because it is not directed and controlled by dual polarity.* The gyroscope and dual polarity of Nature are not a part of the present day laboratory.

A good example is the Haber process of nitrogen fixation which is purely a present day expensive and complex laboratory method of "separating" oxygen and nitrogen, and based upon the belief that each are different substances.

A slight readjustment of Nature's gyroscope will produce nitrogen instead of oxygen—or vice versa. Oxygen is nitrogen divided, and the polarity controlled electric gyroscope is the dividing instrument.

Instead of the expensive and time consuming chemical method of obtaining free nitrogen in LIMITED quantities, Nature's method would produce free nitrogen cheaply, quickly and in UNLIMITED quantities. It is not necessary to call attention to the value to commerce and to agriculture, not to mention soil regeneration, that this method of obtaining nitrogen would be to the world.

In September, 1927 I demonstrated this principle of dual polarity control by arranging two pairs of solenoids—one pair with more windings than the other—in such a manner that the dual polarity of Nature was simulated.

With a steel or glass disc for an equator and a steel rod for amplitude, I adjusted my solenoids approximately to a plane angle where I roughly calculated oxygen belonged in its octave. I improvised an adjustment apparatus which would enable me to fasten any adjustment securely at any angle I chose.

I then inserted a few cubic centimeters of water in an evacuated quartz tube which had electrodes at each end for spectrum analysis readings.

Upon heating the tube in an electric furnace, and inserting it red hot into the solenoid with the electric current turned on until the tube cooled, *the first spectrum analysis showed over 80% to be hydrogen and the rest practically all helium. There was very little oxygen.*

Each time I reset it I obtained a new analsyis. Whenever I set it so the north-south polarity was predominent because of using the stronger coils, the result gave more nitrogen. This was because the preponderant north-south polarity prolated the oxygen atom nucleus to its next higher tone.

When I reversed the polarity to east-west preponderance the analysis showed more than its proper amount of oxygen and inert gases and less of hydrogen. This meant that preponderant east-west polarity had oblated the hydrogen nucleus.

The following analysis is a good example. When I took the tube to the laboratory there was no water in it. That is why the analyst referred to his report as "gas sampleNo. 5," which follows:

Oxygen 14.9
Hydrogen 16.0
Nitrogen or inert gases 69.1

It is needless to say that the above analysis shows east-west preponderance.

I am convinced that by proper adjustments mathematically worked out into formulas by experiment, free hydrogen, nitrogen or oxygen could be obtained without any trace of the others.

The only difference between the two methods of working is that electricity is used as power in the laboratory without polarity control or gyroscopic guidance such as I made use of.

When the gases have been sufficiently transformed by practice the transformation of dense matter can then follow.

XXXXII.
THE AGE OF TRANSMUTATION
NEW CONCEPTS FOR SCIENCE AND NEW VALUES
FOR HUMANITY.

Man must be transformed or perish. Old concepts and old material values must become as obsolete as horse and wagon transportation became obsolete when motors and planes appeared.

Man is still barbarian. Just so long as man kills man he is barbarian. The dawn of his Consciousness is barely six thousand years back in his history. Man must have new concepts, new ideals and new values which will uplift him from the barbarian desires to kill for greed—to build empires for power—to seek happiness through material possessions or to accumulate gold under the delusion that he is creating wealth.

Material values as standards of wealth must be rendered value less. Science has the power to make the transition so gradually that the readjustment will create no hardship to commercial interests and world economy. Just as the transition into the machine age lessened the burdens of man and added to his wealth, so will the transition into the Age of Transmutation have a similar beneficial effect.

All great world transitions which have brought greater ease and wealth to man have been anticipated as calamities. This greatest of all transitions now dawning in man's history, should be looked forward to as the ultimate goal for a peaceful and prosperous unified world.

Man's assets of this age are material. Transformed man must gradually discover that his greatest asset is man. His happiness, achievement and greatest source of wealth and power is in his ability to serve man. The greater his service to man the more he adds to his wealth—both materially and spiritually. For this is LAW—irrevocable and inevitable LAW.

It is inviolate law throughout Nature everywhere. Nature creates its wealth by extending itself into the whole universe from every point in it.

The jungle is rich because it extends all that it has to all of the jungle—while the desert is poor indeed for keeping that which it has within itself. The desert gives naught to the desert, nor to the heavens —therefore its regivings from the heavens are naught.

Nature has no motive for its givings, for regivings are the fulfillment of the law, and man need give no thought to them.

The wealthiest men in all the world are the geniuses who have extended their immortality to other men without thought of gain. These immortals shall never perish from the memory of man who has found his own immortality through them, while he whose wealth is but gold, e'en though it be higher than the highest mountain, shall be forgotten before another dawn.

Man is man's greatest asset therefore, for man's greatest need is other men to whom he may give of his own abundant Self to thus enrich himself through their regivings.

Nature is based upon the Law of Love, which is balanced GIVING for REGIVING. All that man ever has is that which he has given. *That is Nature's only law—and it must eventually become man's only law.*

Nature regives in kind for all service given. Man gives the seed —-and his service in sowing the seed. Nature regives the fruit of the seed. That is Nature's Law. Action is man's free will right but the reaction is Nature's. *It regives equally in kind.*

If man takes a throne it is taken away from him and he is poor indeed. *But if a man enthrones other men, or honors other men, he will be enthroned and honored by other men.*

Spiritual values can replace material ones only by shearing material values of their power to nurture greed and avarice. It will be a slow process but will surely come to pass as science gains the power to shear values from physical matter.

Science has given man this new electric, radio, radar, television age which has made miracles of past-age thinking commonplace today. Had the Nazarine stated that the time would come when the whole world could hear a man's voice He would have then been put to death. Many since then have been burned at the stake and tortured mercilessly for what a schoolboy of today would basically comprehend as Natural Law.

The telephone, the automobile, flight by air, radio, radar and television have been given to the world by science in less than a century. Each of these have transformed man's thinking and his ways of life, for heavy burdens of labor and household drudgery have been lessened for man and woman alike.

The tragic question now arises as to whether the transformations which have affected man's thinking for many centuries have been in the right direction. Are we setting too great a value upon lessened drudgery, greater comforts and other physical values which have multiplied time for man and made this planet very small?

Is our thinking of today right thinking? Can we rightly say it is in the face of the fact that the human race has fallen farther in the last fifty years than it has arisen during seven centuries of forward growth?

Can we say that world thinking of today is right thinking in the face of the undisputable fact that the world is facing a threatened plunge into another period of dark ages?

Have the great scientific contributions of the last hundred years really majored in benefitting the human race by adding to man's comforts and power of production in the direction of peaceful living?

Have the arts of peace been multiplied? Are we producing men of genius in the arts and philosophies such as have enriched the world ever since the days of Angelo, Da Vinci, Mozart, or Shakespeare?

Have our statesmen of the last generation had the moral character, dignity or patriotism of Washington, Jefferson, Lincoln or Theodore Roosevelt?

Have we not found treason replacing patriotism and statesmen more concerned about how to increasingly enslave man and confiscate his earnings to build giant troughs for wastrels gorging?

Has science unwittingly helped to degrade the entire human race by multiplying the arts of war to multiply man's greed for empires by multiplied power to kill? Have these dreadful contributions of science to war not so thoroughly outweighed its contributions to peace that it might not have been better if the bow and arrow days were still here?

What is the responsibility of science in this respect? And can science reverse the results which have grown out of its explosives made to kill men, and save the race by reversing man's thinking?

I think it can—and that is why both my wife, Lao, and I have so indefatigably been working to give this new knowledge to science NOW, when the world is threatened with destruction. This knowledge will enable science to have such command over matter that it will render man's multiplied killing power for greed impotent, and then render those attributes of greed in man also impotent by replacing them with new and greater values in his thinking.

A transformed science can avert this danger which man is bringing upon himself by his own profligacy by rendering all of the coal, oil, nitrates and phosphates of the world not worth the digging for man of today, and needless for man of tomorrow.

These things he can do NOW for they are simple in principle and the means of producing them are simple. A generation need not pass without extending that principle to the heavy metals, and render the gold which constitutes man's idea of wealth—for which he has killed untold millions of his brother men—of no value other than as a utility.

Every product of Nature in the elements of matter which Nature has produced so meagerly can be produced by man in unlimited quantities with less effort than present day digging. This includes iron, copper, manganese, platinum, aluminum, tin, and all other metals. *Man has been primitive long enough. It is time he came into his heritage of knowledge which will give him dominion over the earth.*

In your grind stone is aluminum, and in silicon and carbon—the most plentiful elements of earth—are all of the metals.

Where copper or iron ore cannot be found silicon can give them to us. If we cannot get supplies of tin or manganese from other parts of the world, silicon will give them to us.

The science of metallurgy must realize that all metals are unbalanced conditions of carbon and silicon. Iron and nickel are unbalanced positive and negative extensions of silicon. Silicon is their fulcrum of balance, just as two children on opposite ends of a seesaw are unbalanced extensions of the fulcrum which controls their balance. We now obtain nickel from other countries. *We have an unlimited supply of it in our every mountain.*

Just as sodium and chlorine find balance in their salt, and thus lose their metallic qualities, so do all pairs of metals lose their metallic unbalance in their salts. A salt in Nature is a balanced pair of elements.

Reciprocative balanced reversals of motion is the only power Nature or man has ever used. That is the basis of the electric current—the piston of Nature's wave engines or of man's motors and pumps.

Science has heretofore used but one half of Nature's power principle, and has used even that the hard way. The *easy* way—and the *simple* way— is to use in full the balanced reciprocative reversals of this two-way universe which are forever taking place between the two conditions of gravitation and radiation which motivates this universe.

World strategy of today is largely based upon the location of oil. The world stands ready to kill to protect its supply of oil for fuelling its planes and war ships.

Science can likewise render the oil supplies of the world useless as

*a fuel, and not worth the slightest quarrel among men for the supply
needed for lubrication.*

*But greater than all these is the power of science to verify God
and validate His inviolate Law which gives to man in kind that which
man gives to other men, and thus bring humanity to the realization that
he who would hurt another hurts but himself.*

It is the responsibility of science to unify man's many religions by
giving him full comprehension of the One God of Light and Love to
replace the many ill-conceived imaginings of an impossible God of fear,
which have so disastrously disunited spiritual seekers and divided the
whole world into intolerant and antagonistic groups.

The human race can never become united as one harmonious
whole so long as wrong conceptions of God disunite and divide the
race. Chief among these wrong conceptions is the vengeful God of fear
and wrath which is mainly responsible for the fear, greed, hatred,
superstition and intolerance upon which our present civilization is
based.

*The time has come when science should so inculcate mankind with
the balanced interchange principle of love upon which the universe
is founded, and everywhere manifested in Nature, that the nations of
the earth will become God-loving, instead of God-fearing men.*

Fear of a wrathful God is an inheritance of the terrors of ignorance in primitive man who saw vengeance and wrath of God in the
furies of earth's storms.

Ignorance and terror are still breeding the fears which underlie
our whole world-civilization. World leaders of great vision in science
and government are now the world's great need.

*It may be that our Father in heaven has sent a saviour to end this
greatest of all threats to our American way of life. Our new President
has it in his power—with the aid of science—to perform the miracle of
ending all warfare as past rulers were aided by science to make wars.*

Can this miracle come to pass! Can the thinking of science be
transformed! I think it can but only by being enabled to look at the
world picture of today squarely in the face—see it as it is—and meet it
with new transforming knowledge and the mighty power which attends
transcendent knowledge.

In thus looking squarely in the face of the world today we see God
being driven out of it to deify man. Half the world is drawing an iron

curtain around itself to shut God out and exalt a monster in His place
to dehumanize and enslave man.

We see degradation, corruption, greed, fear, lust for power, and
atheism engulfing half the world, and the tortures of the Russian in-
quisition far exceeding the tortures of the Middle Age Spanish inquisi-
tion as the fruits of today's world thinking.

We see peace, happiness, security and freedom going out of the
world and war engulfing it to enslave and degrade man.

We see beauty and culture being driven out of the world, and the
spiritual rhythms of the fine arts lost in the sea of ugliness which is de-
basing the culture of the race.

We see genius being driven from the face of the earth for want of
recognition and the patronage which alone will nourish it to survival.

We see the swing of the cosmic pendulum away from the glory of
the seven renaissant centuries to another decadent age of forgetful-
ness of all that is good in man.

XXXXIII.
WHAT OF TOMORROW?

Yes—what of tomorrow! We of today are fast using up the resources which have taken the earth millions of years to store up for
man's use. The coal, oil, nitrates and minerals which mankind has taken
from the ground in a hundred years have made big inroads into its total
deposits.

Let us assume that we have five centuries of supply, or even ten
centuries. Man will live on this planet for millions of years before it
spirals out beyond Mars orbit where human life will cease. What of
them?

Are we despoiling the earth for our children of the far tomorrow?
Are we emptying its bins for them? Are we profligately robbing even
the fertility of our soil and losing it into the sea by robbing the mountain sides of their forests?

The greed and ignorance of a few generations of today can wipe
the human race from the face of the earth for long aeons by sheer
wastefulness of earth's resources. It would take millions of years for
Nature to restore balance by bringing continents with new resources
above the seas and taking old and worn out continents under her seas
for regeneration.

WHY ARE WE HERE?

The sole *purpose* of man on earth is to manifest his Creator. He has no other purpose.

The soul *desire* of man on earth is to find peace and happiness.

The only way that man can find peace and happiness is to discover his unity with his Creator. The greatest miracle which can happen to any man is the discovery of his Self, and his oneness with all other men.

To him who has made that supreme discovery all else shall be added.

Knowledge alone will lead man to that supreme discovery. It is the office and responsibility of science to illumine the way for all men who are seeking the kingdom of heaven.

EPILOGUE

By Lao Russell

"All men will come to me in due time but theirs is the agony of awaiting." Thus saith God in His Message of The Divine Iliad.

All down the ages suffering man has lifted up his voice unto his God saying: "Lead us out of the dark of our iniquities into the Light of Thy kingdom.

And God has answered man's prayers through inspired messengers who bring new knowledge of the Light of Love and the Brotherhood of Man into the world for the renewing of man's Mind with the power of new knowing.

But man did not hear God's Voice through His messengers, for man was still new in his primate days of little comprehension. Man crucified God's messengers and again suffered the fall of civilization after civilization by making every man fear every other man.

And yet again in our day the agonies of ten times ten million suffering mothers of men are crying unto God to save the world from another plunge into long ages of darkness. *For once again the human race is nearing another downfall into ages of darkness of its own making, for once again man has made a world of hate where every man fears every other man.*

Over and over again man has climbed far into the heavens in his search for the peace and happiness which Love of man for man alone can give to him, and over and over again he has fallen because he has learned only to hate and fear and kill his fellow man for selfish greed, thinking thus by the power of might he will gain the riches of his seeking.

Man has never known Love as the very heartbeat of this universe —the motivative force behind all matter and motion which controls the stars in their orbits and brings forth the fruits of the earth for man's sustenance.

He has never known Love as Law—irrevocable Law—not emotion or sentiment within man's free will right of giving or taking— but inviolate Law which brings an unescapable penalty to any man who violates that Law in his relations with other men, or with his own body.

He has never known that Love is balanced giving for regiving which Nature obeys in all of its transactions. Man has always taken what he wants, not knowing that the hurt of such taking is his alone.

Man has never known that Love is balanced interchange between the pairs of opposites of this divided universe. Without balance in Nature's transactions the universe could not survive. Likewise without balance in man's transactions man cannot survive.

There never has been balance in man's relation to man. Love has not yet entered the world or the Consciousness of man. Man has never practiced the principle of universal brotherhood which God's messengers gave to age after age of fearing man. There never has been a time in world history when man has not feared and hated his fellow man, and locked his doors and policed his streets because he feared his neighbor.

Nor has there ever been a time when nations of men have not armed themselves in fear of other nations, nor killed when one nation wanted the possessions of other nations or to enslave their peoples for greed of power and gold.

There has never been a time in the blackest day of world history than the black hopelessness of today's world of fear and hate of one half of the world for the other half, and the growing degradation and lowering of the spiritual standards of all the world.

This disunited, fear ridden, tax burdened world of man's centuries of empire building by conquest of the weak by the strong cannot survive. It is doomed to self-destruction unless at this eleventh hour the lesson of Love, once again given to man in God's Divine Iliad Message, is learned and heeded by the few among men to whom God will give new power to immunize the few from the harm of the many.

Unless the few among the leaders of men will arise to the power of new knowing given in God's Message of The Divine Iliad the free world of man will disappear. The slave world will then appear as a foremath to unthinkable degradation of the whole human race.

Where Love is there also is unity, harmony and the peace of Love's balanced rhythms in a united world. Where hate is there follows the degeneracy of disunity as night follows the day.

That is the lesson which unfolding man has still to learn. *Until he learns that simple lesson of power which comes from giving of service to his fellow man, instead of taking from him against his will, his civilizations will disappear in their own man-made chaos, one after another, until he learns that lesson.*

After millions of years of *taking* by the power of his might, his six thousand years out of the jungle have not been long enough for him to learn that lesson of power which lies alone in the *giving* of Love, nor has he yet learned that his destruction is of his own making through violation of the Law of Love.

Man acquired no knowledge and but little comprehension during his slow unfolding through primate and pagan ages, for he was not ready for it. Consciousness of Mind in him had not yet dawned. Through dense ignorance of God's ways man has suffered the agonies caused by dense ignorance.

Then came the dawn of Consciousness in barbarian man and his first suspicions of a God-Creator who to him was a vengeful God of wrath for whom he shed the blood of bullocks and even men upon sacrificial altars to appease his vengeful God of fear and wrath.

God sent new knowledge and His message of Love and the unity of man through illumined messenger after illumined messenger all through his early barbarian days, but man was still too new to comprehend, for he was not yet ready to comprehend a God of Love nor His message of Love. He still shed blood upon his sacrificial altars to appease his vengeful God of fear, and he still suffered the agonies of his little comprehension.

Man is still barbarian, for man still kills man; and he still worships a wrathful God of fear. And man will forever suffer the agonies of his ignorance until Mind awareness of the God of Love awakens in him in its fullness, and man knows man as one brotherhood, and begins to serve man instead of killing him.

Man learns his lessons by deep suffering, for only at times of great suffering does he turn to God for Light to illumine his path out of his dark pit of hopeless despair.

Man of today has had a half century of deep suffering and many there are among men who have turned their faces to the high heavens and cried aloud to be saved from their agonies.

Man of today is not so new. His comprehension is now great enough to understand God's ways as manifested in His one Law of Love. Man of today is ready for new knowledge and God has given to those few who are able to comprehend it the power of new knowing to command the forces which order the movements of stars in their orbits and the earth to bring forth its fruits.

The knowledge of God's ways given to man for his new day will give the few among men mighty new power to control all men of earth through God's One Law of Love until the seed of it will multiply over the face of the earth and bring with it the harmony and peace of its balanced rhythms.

When Mind Consciousness dawns in man God awareness likewise dawns in him, and he becomes illumined with full knowing of the Oneness of Mind of man and Mind of God.

When that day dawns for man he has command over all the universe, for energy of Mind in him created the universe, and knowledge of Mind in Him controls its energy.

Fear then leaves him, for he knows he has dominion over all things. He can no longer be hurt by man, nor will he hurt man: but the power will be his to prevent man from hurting man by awakening Consciousness in him to Love, e'en though he may lose one more life to find it.

God's one Message of Love—which he again sends to man for his new day—is written down in The Divine Iliad Message in the following imperishable words of man's understanding:

"Great art is simple. My universe is great art, for it is simple.

"Great art is balanced. My universe is consummate art, for it is balanced simplicity.

"My universe is one in which many things have majestic measure: and again another many have measure too fine for sensing.

"Yet have I not one law for majestic things, and another law for things which are beyond the sensing.

"I have but one law for all my opposed pairs of creating things: and that law needs but one word to spell it out, so hear me when I say that the one word of My one law is

BALANCE

"And if man needs two words to aid him in his knowing of the workings of that law, let those two words be

BALANCED INTERCHANGE

"If man still needs more words to aid his knowing of My one law, give to him another one, and let those three words be

RHYTHMIC BALANCED INTERCHANGE."

—*From THE DIVINE ILIAD*

He who reads these words with inner vision and inner knowing shall have omnipotent power to save the world of man from himself and bring into being the new age of man's new power.

Portfolio of Explanatory Diagrams

Reproduced from

The Home Study Course

of the

RUSSELL COSMOGONY

by Walter and Lao Russell

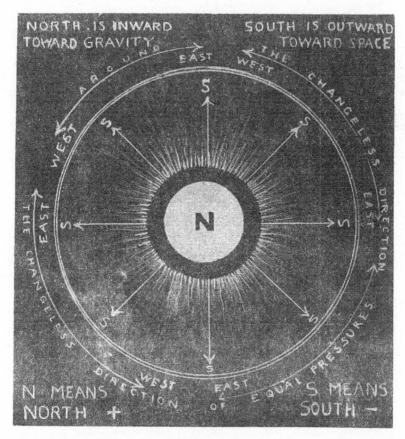

Fig. 16

THE TWO DYNAMIC DIRECTIONS OF CHANGING PRESSURES, AND THE ONE
STATIC, CHANGELESS DIRECTION OF EQUAL PRESSURES

EVERY CHANGING EFFECT IN NATURE PULLS INWARD FROM WITHIN TO BUILD
BODIES, AND THRUSTS OUTWARD FROM WITHIN TO DESTROY THEM

The inward direction of gravitation compresses light waves of
matter into incandescent spheres of high potential. The outward
thrust of radiation expand light waves into low potential gases
and ethers of cold, dark space to surround the solid spheres.

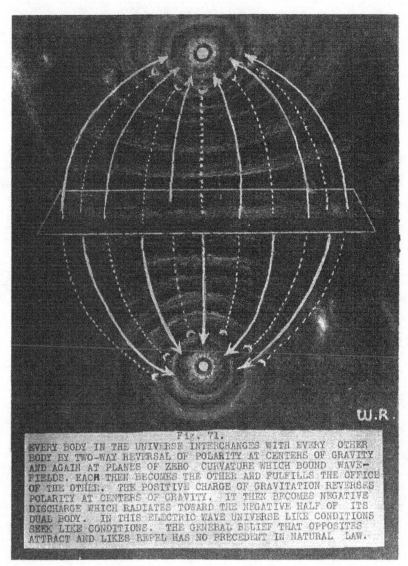

Fig. 71.

EVERY BODY IN THE UNIVERSE INTERCHANGES WITH EVERY OTHER
BODY BY TWO-WAY REVERSAL OF POLARITY AT CENTERS OF GRAVITY
AND AGAIN AT PLANES OF ZERO CURVATURE WHICH BOUND WAVE-
FIELDS. EACH THEN BECOMES THE OTHER AND FULFILLS THE OFFICE
OF THE OTHER. THE POSITIVE CHARGE OF GRAVITATION REVERSES
POLARITY AT CENTERS OF GRAVITY. IT THEN BECOMES NEGATIVE
DISCHARGE WHICH RADIATES TOWARD THE NEGATIVE HALF OF ITS
DUAL BODY. IN THIS ELECTRIC WAVE UNIVERSE LIKE CONDITIONS
SEEK LIKE CONDITIONS. THE GENERAL BELIEF THAT OPPOSITES
ATTRACT AND LIKES REPEL HAS NO PRECEDENT IN NATURAL LAW.

EVERY ANODE IS ALSO A CATHODE AND EVERY CATHODE IS AN ANODE.
EVERY CHARGING BODY IS ALSO DISCHARGING,AND EVERY DISCHARGING
BODY IS ALSO CHARGING. IN THIS MANNER LIFE GIVES TO DEATH
THAT DEATH MAY DIE,AND DEATH GIVES TO LIFE THAT LIFE MAY LIVE.

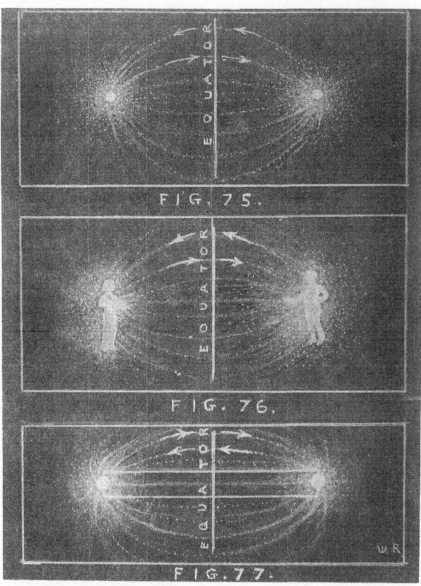

FIG. 75.

FIG. 76.

FIG. 77.

GOD   IS   LIGHT ---- AND   GOD   IS   LOVE
LOVE IS DIVIDED INTO PAIRS OF OPPOSITES — GIVING — AND REGIVING
RADIATION IS THE ACTION OF GIVING. ITS REACTION IS GRAVITATION.

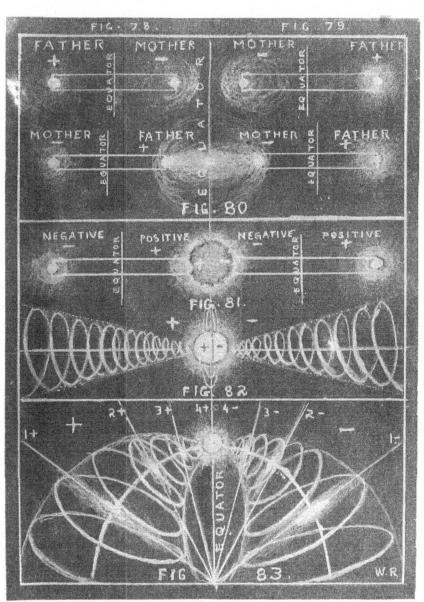

ILLUSTRATING THE FATHER-MOTHER PRINCIPLE OF BUILDING BODIES BY
DIVIDING LIGHT INTO POLARIZED UNITS--AND REPRODUCING BODIES BY
UNITING TWO WAVE-FIELDS OF OPPOSED UNITS INTO ONE. -- SEE TEXT.

FIG. 85.

IN THE OCTAVE LIGHT WAVE LIES THE SECRET OF CREATION AND ALL OF
ITS PROCESSES. AT THE LEFT ARE THE POLARIZED MOTHER-FATHER UNITS
AND IN THE CENTER ARE FATHER-MOTHER HALVES UNITED IN MARRIAGE TO
PRODUCE A PERFECTLY BALANCED SPHERICAL BODY.

FIG. 86.

CROSS SECTION OF OCTAVE WAVE ILLUSTRATING THE GYROSCOPIC SPIRAL
PRINCIPLE OF MULTIPLICATION OF POWER BY ACCELERATING SPEED CEN-
TRIPETALLY TO BUILD A SPHERE, THEN DIVIDING POWER BY DECELERAT-
ING SPEED CENTRIFUGALLY UNTIL MOTION IS AGAIN ZERO AT WAVE AXIS.

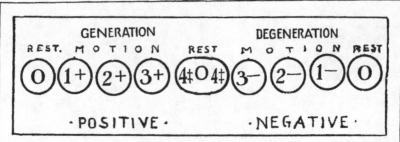

Fig. 87.

Octave Waves of Vibrating Light, Which Constitute the Elements of Matter, Consist of Four Pairs of Tones, Each Centered by a Zero Fulcrum of Rest Which Controls Their Balance From Within. Two Pairs of Magnetic Poles Control Their Balance From Without.

When Any of These Two Equal Pairs of Opposite Tones Are United They Become Stable, Such as Sodium Chloride. When They Become Disunited They Are Then Unstable, Such As Sodium and Chlorine.

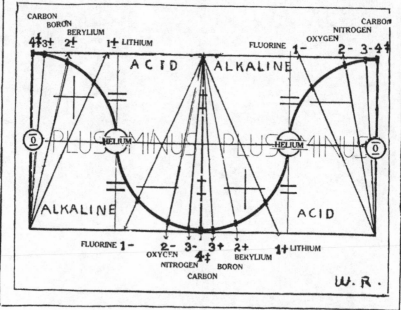

Fig. 88.
SHOWING ONE OF THE NINE OCTAVES OF THE ELEMENTS OF MATTER.

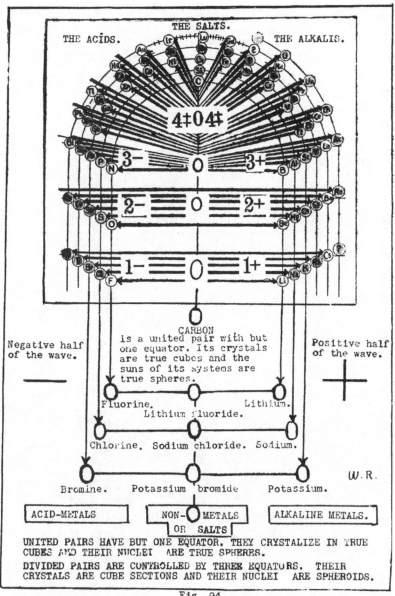

Fig. 94.
IN THE ABOVE FIVE, OF THE NINE OCTAVES OF MATTER, THE FULCRUM AND
LEVER PRINCIPLE OF DIVIDING AND EXTENDING ONE BALANCED CONDITION
INTO TWO OPPOSED CONDITIONS TO MOTIVATE THE HEARTBEAT OF NATURE
IS GRAPHICALLY ILLUSTRATED.

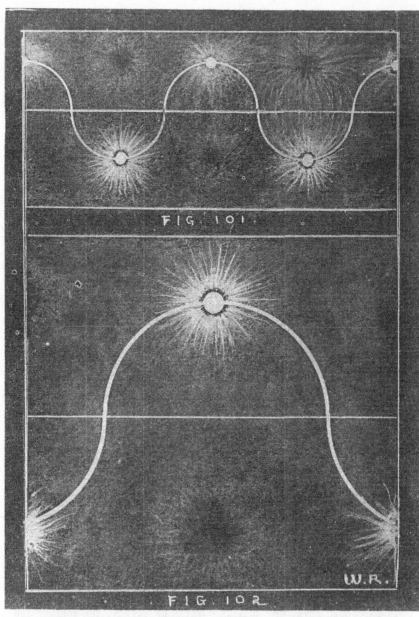

FIG. 101

FIG. 102

ANY FORCE EXPRESSED ANYWHERE SIMULTANEOUSLY CREATES AN EQUAL
AND OPPOSITE FORCE IN REVERSE TO VOID IT--AND TO REPEAT IT.

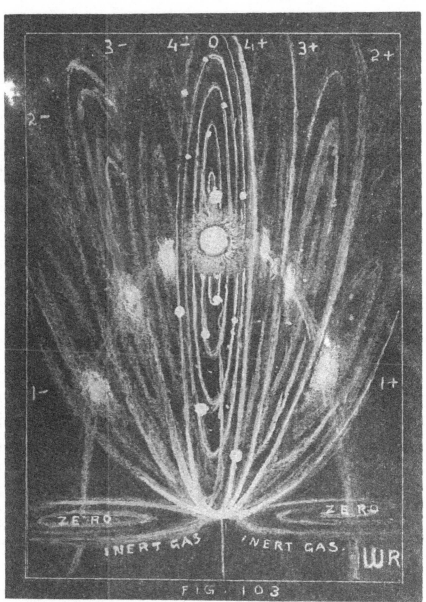

FIG. 103

ANY SYSTEM, WHETHER OF AN ATOM OF THE ELEMENTS OR A SOLAR SYSTEM
IN OUR MILKY WAY, IF UPON THE AMPLITUDE OF ITS WAVE AND NINETY
DEGREES FROM ITS WAVE AXIS, ITS CENTRAL SUN IS A TRUE SPHERE AND
ALL OF ITS PLANETS REVOLVE UPON THE PLANE OF THEIR SUN'S EQUATOR.

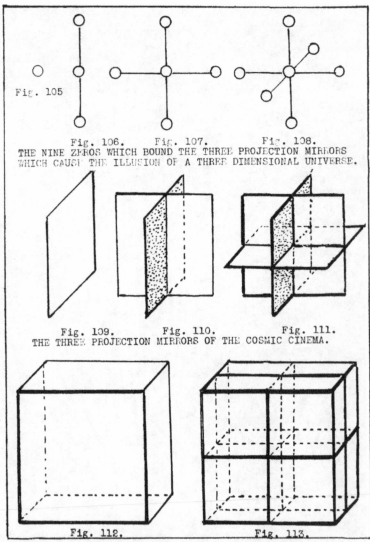

Fig. 105

Fig. 106.   Fig. 107.   Fig. 108.
THE NINE ZEROS WHICH BOUND THE THREE PROJECTION MIRRORS
WHICH CAUSE THE ILLUSION OF A THREE DIMENSIONAL UNIVERSE.

Fig. 109.   Fig. 110.   Fig. 111.
THE THREE PROJECTION MIRRORS OF THE COSMIC CINEMA.

Fig. 112.   Fig. 113.

FIG. 112 -- THE SIX MIRRORS WHICH FORM THE SCREEN OF SPACE
UPON WHICH THE COSMIC DRAMA OF CREATION IS THROWN.
FIG. 113 -- THE WHOLE WAVE-FIELD PROJECTION MACHINE OF NINE
MIRRORS WHICH CREATE THE ILLUSION OF FORM AND MOTION IN A
ZERO UNIVERSE OF ELECTRICALLY RECORDED MIND IMAGININGS.

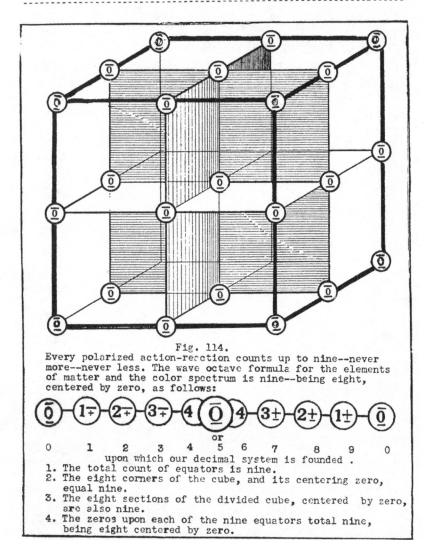

Fig. 114.
Every polarized action-reaction counts up to nine--never
more--never less. The wave octave formula for the elements
of matter and the color spectrum is nine--being eight,
centered by zero, as follows:

$$\overline{\underline{0}} - 1\tfrac{+}{} - 2\tfrac{+}{} - 3\tfrac{+}{} - 4 \; \overline{\underline{0}} \; 4 - 3\pm - 2\pm - 1\pm - \overline{\underline{0}}$$

or

| 0 | 1 | 2 | 3 | 4 | 5 | 6 | 7 | 8 | 9 | 0 |

upon which our decimal system is founded .

1. The total count of equators is nine.
2. The eight corners of the cube, and its centering zero,
   equal nine.
3. The eight sections of the divided cube, centered by zero,
   are also nine.
4. The zeros upon each of the nine equators total nine,
   being eight centered by zero.

THE UNIVERSAL EQUILIBRIUM IS PROTECTED FROM BEING UPSET BY WAVE-
FIELD SYSTEMS OF NINE EQUATORS. WITHIN THESE INSULATING FIELDS
POLARIZATION CAN EXPRESS ITS OPPOSITION, BUT CANNOT PASS BEYOND.

Size of sphere many millions of times exaggerated.

FIG. 115   W.R.

SPHERES REACH COMPLETION INTO TRUE SYMMETRICALLY BALANCED FORMS ONLY
AT WAVE AMPLITUDES WHERE DIAGONALS OF EIGHT WAVE-FIELDS MEET. FOR
THIS REASON SPHERES HAVE WITHIN THEM THE EIGHT OCTAVE TONES WHICH THEY
PROJECT RADIALLY INTO THE EIGHT SEPARATE OCTAVE WAVE TONES OF THE
SPHERE WHICH IS FATHER-MOTHER OF THE OTHER THREE DIVIDED PAIRS .

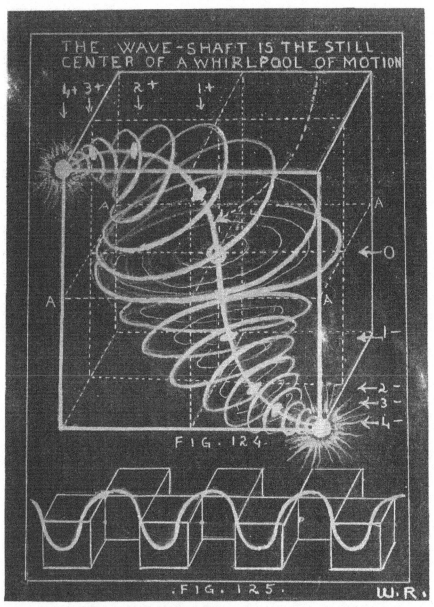

ILLUSTRATING NATURE'S METHOD OF WINDING LIGHT UP INTO SOLID SPHERES TO
CREATE THE CONDITION OF GRAVITY AT TROUGHS AND CRESTS OF WAVES. THE
EFFECT OF GRAVITATION IS PRODUCED BY THE INWARD PULL OF CENTRIPETAL FORCE.

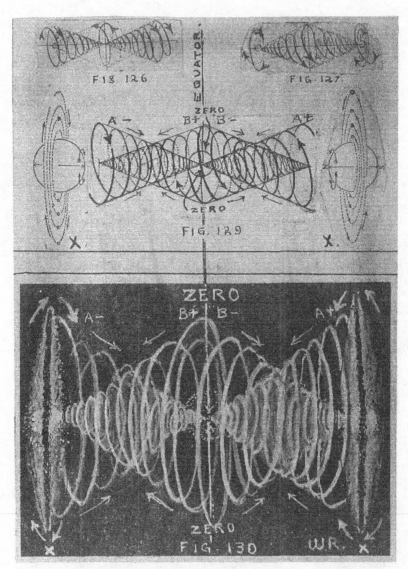

LOOPS OF FORCE IN AN ELECTRIC CURRENT ARE WOUND UP CENTRIPETALLY
JUST AS SOLAR AND STELLAR SYSTEMS ARE WOUND UP IN THE HEAVENS.
CENTRIFUGAL FORCE UNWINDS THEM FOR REWINDING AND REPETITION.

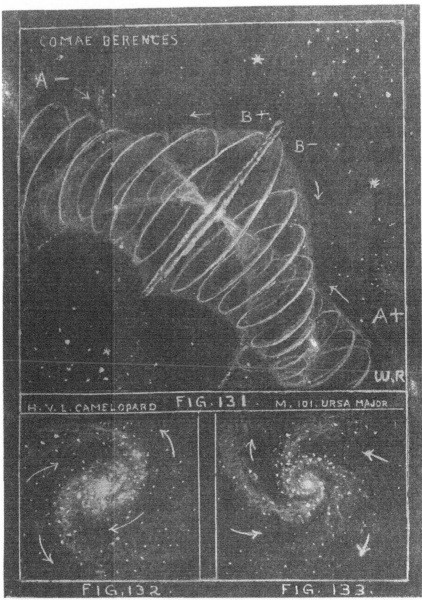

FIG. 131 ILLUSTRATES METHOD OF CREATING INCANDESCENCE BY MULTIPLYING
DARKNESS. INCANDESCENCE IS THEN DIVIDED TO AGAIN BECOME DARKNESS.
FIG. 132 AND 133 SHOW GIANT NEBULAS COMPRESSING DARKNESS INTO LIGHT,
AND EXPANDING LIGHT INTO DARKNESS TO CREATE BODIES AND DESTROY THEM.

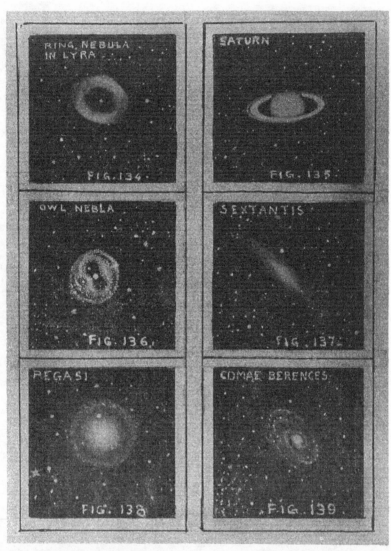

ELECTRICITY DISINTEGRATES SPHERES BY UNWINDING THEM BY THE WAY OF
THEIR EQUATORS. GREAT RINGS ARE THROWN OFF WHICH REWIND TO BECOME
SATELLITES. THIS PROCESS CONTINUES UNTIL MATTER EXPANDS INTO ZERO.

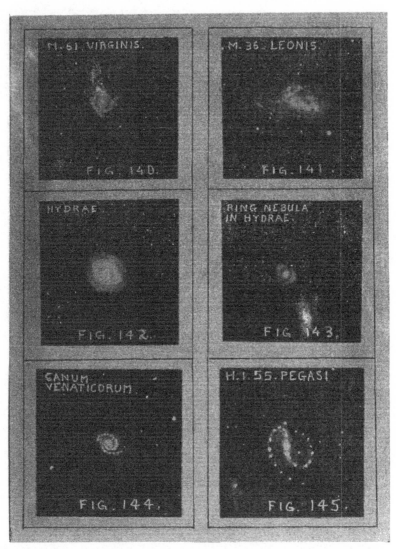

ELECTRICITY INTEGRATES SPHERES BY WINDING LIGHT AROUND THEIR POLES
OF ROTATION.   THE CENTRIPETAL FORCE OF GRAVITY WINDS THEM INTO
HIGH POTENTIAL CENTERS OF GYROSCOPIC SYSTEMS OF LESSER SPHERES.

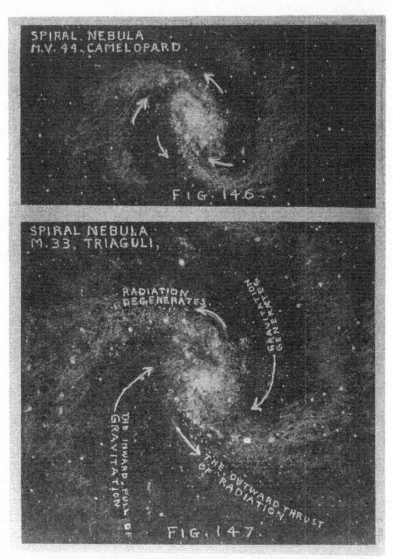

NOTE THE DIFFERENCE IN BALANCE BETWEEN THE TWO POLARIZING
FORCES IN THESE TWO PAST MIDDLE AGE NEBULAS. THE LOWER ONE
INDICATES PERFECT BALANCE BETWEEN THE TWO FORCES WHICH
CREATED IT. ARROWS POINT TOWARD BOTH OPPOSING DIRECTIONS.

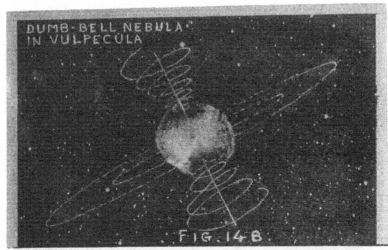

THE DEATH OF A SYSTEM BY EXPANSION. CENTRIFUGAL FORCE HAS BORED A
HOLE THROUGH THIS ONCE INCANDESCENT SUN AND MADE A RING OF IT. A
SMALLER SUN IS FORMING AT ITS CENTER SUCH AS IN FIGURES 134 & 136.

THE BIRTH OF A SYSTEM BY CONTRACTION. BILLIONS OF YEARS FROM NOW
THIS NEBULOUS MASS WILL BE WOUND UP TO THE STAGE SHOWN IN FIGURE
154. BILLIONS OF AGES LATER IT WILL MATURE TO THE STAGE SHOWN IN
FIGURES 133 & 147. IT WILL STILL TAKE UNTOLD AGES TO REACH ZERO.

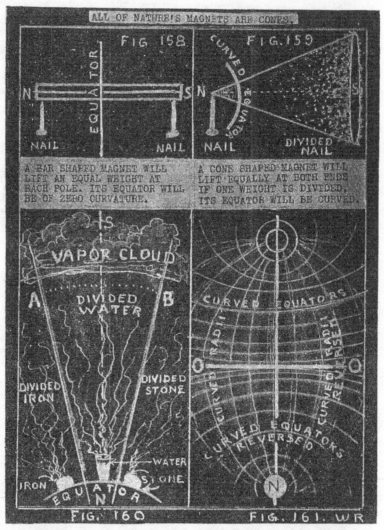

ALL OF NATURE'S MAGNETS ARE CONES.

FIG. 158

A BAR SHAPED MAGNET WILL LIFT AN EQUAL WEIGHT AT EACH POLE. ITS EQUATOR WILL BE OF ZERO CURVATURE.

FIG. 159

A CONE SHAPED MAGNET WILL LIFT EQUALLY AT BOTH ENDS IF ONE WEIGHT IS DIVIDED. ITS EQUATOR WILL BE CURVED.

FIG. 160

FIG. 161. WR

SOLIDS ARE COMPRESSED GASES. GASES ARE DIVIDED SOLIDS.

MULTIPLIED MATTER WILL FALL AND DIVIDED MATTER WILL RISE.	RADII CURVE IN THE LENSES OF CURVED EQUATORS.

IN THIS CURVED UNIVERSE GRAVITY AND RADIATION ARE CURVED.
GRAVITY FUSES   RADIATION DIFFUSES.

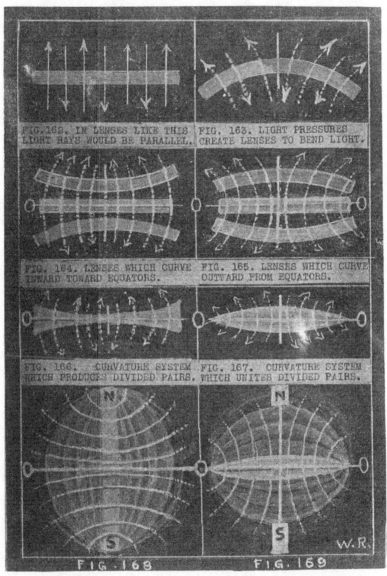

FIG. 162. IN LENSES LIKE THIS   FIG. 163. LIGHT PRESSURES
LIGHT RAYS WOULD BE PARALLEL.   CREATE LENSES TO BEND LIGHT.

FIG. 164. LENSES WHICH CURVE   FIG. 165. LENSES WHICH CURVE
INWARD TOWARD EQUATORS.   OUTWARD FROM EQUATORS.

FIG. 166.   CURVATURE SYSTEM   FIG. 167.   CURVATURE SYSTEM
WHICH PRODUCES DIVIDED PAIRS.   WHICH UNITES DIVIDED PAIRS.

FIG. 168   FIG. 169

ILLUSTRATING NATURE'S METHOD OF PRODUCTION AND REPRODUCTION.
THIS SYSTEM OF CURVATURE   THIS SYSTEM OF CURVATURE
DIVIDES ONE BALANCED   UNITES TWO UNBALANCED
CONDITION INTO TWO UNBAL-   CONDITIONS INTO ONE AND
ANCED CONDITIONS.   MULTIPLIES THEM.

ON THE THREE EQUATORS OF ZERO CURVATURE IN A LIGHT SPHERE LENS ALL RADII LOSE THEIR CURVATURE TO REVERSE IT IN THE EIGHT SECTIONS OF THE SPHERE WHICH THESE EQUATORS DIVIDE.

FIG. 170

ALL OTHER RADII IN A LIGHT SPHERE LENS CURVE INWARD TOWARD POLES TO THE CENTER OF THE LENS TO FORM A CENTER OF GRAVITY WHERE LIGHT COMPRESSION IS MAXIMUM.

FIG. 171

CORONAS BECOME RINGS AND RINGS BECOME SYSTEMS

FIG. 172

THE OUTWARD THRUST OF RADIATION UPON RADII WHICH BEND AWAY FROM POLES TO LOSE THEIR CURVATURE IN EQUATORS AND GAIN IT AGAIN IN REVERSE, TOGETHER WITH THE INWARD PULL OF RADII WHICH BEND TOWARD POLES AND LOSE THEIR CURVATURE IN POLES AND GAIN IT AGAIN IN REVERSE, CAUSES MAXIMUM RADIATION AT EQUATORS AND MAXIMUM GENERATION AT POLES. THUS ARE SUNS WOUND UP AND UNWOUND.

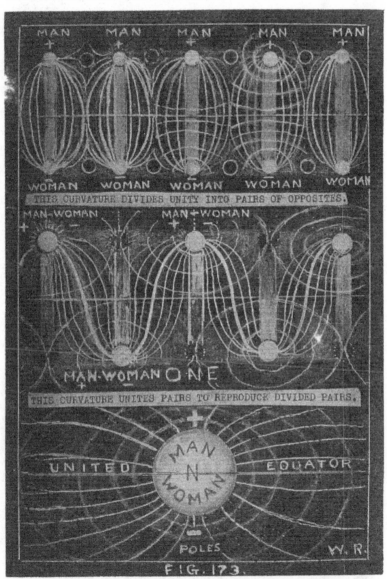

FIG. 173.

NATURE IS FOREVER DIVIDING THE UNITY OF FATHER-MOTHERHOOD INTO
SEX DIVIDED FATHERS AND MOTHERS WHICH UNITE INTO THE ONENESS OF
FATHER-MOTHERHOOD TO MULTIPLY SEX DIVIDED FATHERS AND MOTHERS.

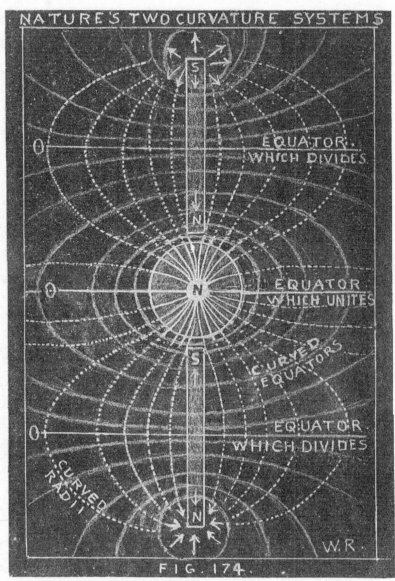

ILLUSTRATING NATURE'S REPRODUCTIVE PROCESS.   CURVATURE FOR
DIVIDED PAIRS MUST REVERSE FOR UNITED PAIRS.   TWO EQUATORS
MUST BECOME ONE AND TWO CENTERS OF GRAVITY MUST ALSO BE ONE.

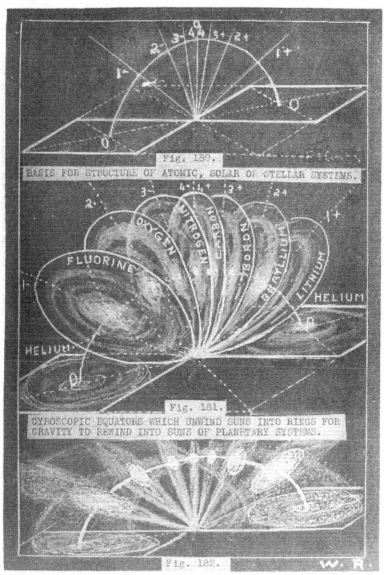

Fig. 180.

BASIS FOR STRUCTURE OF ATOMIC, SOLAR OR STELLAR SYSTEMS.

Fig. 181.

GYROSCOPIC EQUATORS WHICH UNWIND SUNS INTO RINGS FOR GRAVITY TO REWIND INTO SUNS OF PLANETARY SYSTEMS.

Fig. 182.

W. R.

ATOMIC PLANETORY SYSTEMS EXTEND FROM THEIR ZERO INERT GASES BY WINDING SPIRALLY IN A SERIES OF FOUR TONAL VORTICES WHICH ARE RECORDED IN MATTER AS PROLATING SPHEROIDS. THESE BECOME TRUE SPHERES WHERE WAVE MEETS WAVE AT AMPLITUDES.   ACCELERATION OF REVOLUTION WINDS SPHERES, ACCELERATION OF ROTATION UNWINDS THEM.